T0351284

CONSUMERS AND NANOTECHNOLOGY

Second Edition

CONSUMERS AND NANOTECHNOLOGY

Deliberative Processes and Methodologies

edited by
Pål Strandbakken | Gerd Scholl | Eivind Stø

JENNY STANFORD
PUBLISHING

Published by

Jenny Stanford Publishing Pte. Ltd.
Level 34, Centennial Tower
3 Temasek Avenue
Singapore 039190

Email: editorial@jennystanford.com
Web: www.jennystanford.com

British Library Cataloguing-in-Publication Data
A catalogue record for this book is available from the British Library.

Consumers and Nanotechnology: Deliberative Processes and Methodologies

Copyright © 2021 by Jenny Stanford Publishing Pte. Ltd.
All rights reserved. This book, or parts thereof, may not be reproduced in any form or by any means, electronic or mechanical, including photocopying, recording or any information storage and retrieval system now known or to be invented, without written permission from the publisher.

For photocopying of material in this volume, please pay a copying fee through the Copyright Clearance Center, Inc., 222 Rosewood Drive, Danvers, MA 01923, USA. In this case permission to photocopy is not required from the publisher.

ISBN 978-981-4877-61-9 (Hardcover)
ISBN 978-1-003-15985-8 (eBook)

Contents

Preface

The material for the first edition of this book (2013) was mainly based on results from the European Union's 7th Framework Program, Science in Society project *Nanoplat: Development of a Platform for Deliberative Processes on Nanotechnology in the European Consumer Market*. Most chapters started life as project deliverables for the Nanoplat project, but were expanded, re-written and re-edited for the book. The first edition was edited by Pål Strandbakken, Gerd Scholl, and Eivind Stø. The original 'empirical' chapters have been left untouched for this book; documenting nano deliberations' state of art at a specific point in time. The original's Chapter 13, that concluded the first edition, was to a large extent based on work done in a project financed by the Norwegian Research Council called *Deliberative Processes: Increased Citizenship or a Threat to Democracy? A Synthesis of Recent Research Within Nanotechnology*.

The Nanoplat project was coordinated by the National Institute for Consumer Research, SIFO, situated in Oslo, Norway. The other partners were Institut für Ökologische Wirtschaftsforschung, IÖW, Berlin, Germany, the Central European University, CEU, Budapest, Hungary, the Sabanci University, Istanbul, Turkey, Manchester University, Manchester, UK, the Centre for the Study of Science and the Humanities at the University of Bergen, Bergen, Norway and the Strategic Design Solutions, SDS, Brussels, Belgium. All the authors were involved in the Nanoplat project, and at the time they were all affiliated with one of these partners, except for Fern Wickson's co-authors Michael D. Cobb and Patrick Hamlett (Chapter 9).

The Deliberative Processes project was led by SIFO and included IÖW as partner.

Chapter 14, written for this new edition is based on work done in the project: *NanoDiode: Developing Innovative Outreach and Dialogue on Responsible Nanotechnologies in EU Civil Society*, a coordination and support action also under the 7th Framework Program. The project was running from 2013 to 2016 and included 14 partners/beneficiaries from the Netherlands (3), France (3), Belgium (2), Germany, Italy, UK, Austria, Poland, and Norway. The project was led by IVAM UVA in the Netherlands and it focused on stakeholder engagement and dialogue, attempting to contribute to the responsible development of nanotechnologies in Europe.

Chapter 15, also written for this new edition is based on work done in the large European project *Strenght2Food*. It presents a version of Callon *et al.*'s 'Hybrid Forum', called HF 2.0. This was not a project targeting nanotechnology, but the methodological approach and its take on citizen and stakeholder engagement is interesting, and most likely transferable to nanotechnology and other emerging technologies.

The third new contribution is Chapter 16, called 'Conclusion 2020. A more democratic science through public engagement?' The title pretty much sums it up. It is an assessment of the status of deliberations after the 'turn to RRI' post 2010, a comparison with Strenght2Food's version of hybrid forums, called HF 2.0 and a new look at the relation between deliberative, or participatory democracy and the representative system, building on the experience from the analysed cases.

The enlarged and re-edited Chapter 1, on emerging technologies, science studies, engagement and direct democracy is based on work in and around the aforementioned projects (Nanoplat and Deliberative Process: Increased Citizenship or a Threat to Democracy?), plus NanoDiode and Strenght2Food.

Editing, post-production, and development of additional texts and material for the first edition was financed by SIFOs programme for the Study of Material Culture, led by Pål Strandbakken. This remains *a nice reminder that no matter how small nanoparticles are, they are still material objects.* The work on this new edition is financed by SIFO Consumption Research Norway, now a part of Oslo Metropolitan University.

Pål Strandbakken
Oslo and Moss, February 2021

Outline of the Book

Pål Strandbakken

Nanotechnologies are a set of technologies that enable manipulation of structures and processes at the atomic level. It is the ability to do things — measure, see, predict, and make — on the scale of atoms and molecules and exploit the novel properties found at that scale. The nanoscale is not just another step towards miniaturisation, but a qualitative new scale. Today we find some nanoproducts in the consumer market, such as cosmetics, sport equipment, household appliances, and ICT products; but we are at an early stage in the innovation process, and the main results of scientific and industrial investments in nanotechnologies are still to come (Throne-Holst and Stø, 2008).

This is NOT a book about nanotechnology. It is a book about democratising science, using nano as a case. It describes and analyses a set of what we call deliberative processes on nanoscience and nanotechnology. Deliberative processes are exercises in participatory democracy; either for involving citizens or for involving stakeholders, like businesses and NGOs in policymaking. This aim at involving society can appear for a number of reasons; one might be to ease the introduction of a new technology, another to broaden the knowledge base and democratic legitimacy of science policymaking. This list could be extended.

The book is divided into four parts. **Part I, Science and Democracy**, consists of two chapters. The first is a theoretical introduction called *Emerging Technologies and Democracy* (Chapter 1), written by Pål Strandbakken. The chapter aims to put the reviewed processes into a theoretical context, mentioning, but not going deeply into some relevant traditions and debates like Science and Technology Studies (STS), theories of 'new' governance and on participatory democracy.

The second chapter, *Overview of a Set of Deliberative Processes on Nano* (Chapter 2), written by Gerd Scholl and Ulrich Petschow is based on the Nanoplat project's first description of the landscape of nano deliberation, and a selection of candidates for further analysis. Initially, more than 60 processes were collected, in Europe and abroad (abroad mainly meaning the United States). Further, the chapter maps the 'intensity' of the deliberation in a figure using the number of participants and the length of the process as variables.

In **Part II, Citizen-Oriented Deliberative Processes**, we, as the title suggests, review and analyse seven different deliberations, all aiming at involving the common man, the non-expert. The first chapter, written by Eivind Stø, deals with what has been regarded as the first deliberative process on nano ever, the *Citizens' Nano Conference in Denmark* (Chapter 3), arranged in June 2004. This has later come to be known as the 'Danish model' and as a 'consensus conference'. This small, one-evening conference with 29 participants has had an amazing impact on both nano deliberations elsewhere and actually on Danish science policy as well.

Eivind Stø has also written the account of *The NanoJury in the UK* (Chapter 4); a deliberation with fewer participants (only 16) than the Danish conference, but with more time resources, since it lasted for five weeks. Here the jury concept actually implies a break with the idea of a consensus conference. In the jury, disagreement is solved by majority voting.

The third process is the *Consumer Conference on Nanotechnology in Food, Cosmetics and Textiles*, Germany (Chapter 5). The chapter is written by Gerd Scholl. The number of participants is still rather small, 16 again, but here the whole process was longer, lasting up to 10–11 months, the actual deliberation with participants going on for five to six. The panel had good access to experts as well. One notable feature in addition is that the focus on consumption, even on three specific areas of consumption, made this conference less general than most other deliberations. It addressed specific consumer topics, rather than dealing with science and society in the abstract.

With the *French Conference Cycle on Nanotechnology, Nanomonde* (Chapter 6), we meet a different type of deliberation. Giampiero Pitisci's chapter deals with a cycle of six — one per month for six months — large conferences, with as many as 100 to 120 citizen participants. The Nanomonde cycle was arranged by an NGO called Vivagora, as 'vigilance on democracy' activity and aimed at representing the public voice, facilitating an open dialogue between public and experts.

The second French deliberation included here, and the last of the European citizen-oriented processes, is the *Citizens' Conference,*

Île-de-France (Chapter 7); chapter written by Pål Strandbakken. This event is interesting for a number of reasons; it had quite large resources, administratively, expert wise, and duration wise, and it is well documented through a DVD, but it is mainly interesting here because of its close ties to politics and decision-making. It was explicitly arranged to produce input for the science policy decision-making in the Île-de-France region.

The *Policy Recommendations from the Citizens' Conference* is printed as Appendix 1, after the chapter.

The chapter on *Nanotechnology Citizens' Conference in Madison, USA* (Chapter 8) is written by M. Attila Öner and Özcan Saritas. This was a small, university-based consensus conference, modelled on the citizens' nano conference in Denmark, in a different political culture, but with more time at hand. The Madison conference is described as something like an 'ideal deliberation', but it suffers from an apparent lack of political impact.

The *Written Submission from the Citizens' Coalition on Nanotechnology* is printed as Appendix 2, after the chapter.

The final and probably best-documented citizen deliberation is *The U.S. National Citizens' Technology Forum on Human Enhancement: An Experiment in Deliberation Across a Nation* (Chapter 9), written by Fern Wickson, Michael D. Cobb, and Patrick Hamlett. It is interesting in itself, because of the original use of both face-to-face and online deliberations, and it covered a large geographic area. Design wise, the National Citizens' Technology Forum (NCTF) process is interesting because participants were asked to fill in questionnaires before and after the event. Two of the authors (Cobb and Hamlett) were partners in the initiative and evaluated the original process.

The list of facilitators in the six state teams is printed as *Appendix 3.*

Part III, Stakeholder-Oriented Deliberative Processes, consists of three chapters. The first is *Experiments with Cross-National Deliberative Processes Within FP6 and FP7 of the European Union: The Convergence Seminars, the DEMOCS Card Games, and the Nanologue Project* (Chapter 10), by Eniko Demeny, Judith Sandor, and Peter Kakuk. The title pretty much sums it up. The three experiments all clearly go beyond the citizen-oriented processes, involving experts and stakeholders. In addition, all three could be labelled as deliberative processes within research processes.

The second chapter *Standardisation as a Form of Deliberation* (Chapter 11) is written by Harald Throne-Holst and Pål Strandbakken. In it, the authors account for the rules, methods, and ideas/ambitions that drive the standardisation work on nano in the European standardisation body, Comité Européen de Normalisation (CEN), and in the International Organization for Standardization (ISO). The chapter does not analyse a specific deliberative process but rather aims at analysing the conditions for deliberation in nano standardisation processes in general.

In the third chapter, *An Online Platform for Future Deliberative Processes* (Chapter 12), Francois Jegou and Pål Strandbakken account for the development and test runs of Nanoplat's deliberation tool. It was designed as a social computing like tool for consensus conferences among stakeholders.

Part IV, Methods and Approaches for Stakeholder and Citizen Involvement, consists of four chapters. The first, *Conclusions: Towards a Third Generation of Deliberative Processes* (Chapter 13) is written by the editors Eivind Stø, Gerd Scholl, and Pål Strandbakken. It concludes the original edition and elaborates on the idea of different generations of deliberative processes, arguing for a development towards more specific themes and clearer political linkages of a proposed future 'third generation of deliberative processes on nanotechnology'.

The second chapter, *Third Generation Deliberative Processes on Nanotechnology* (Chapter 14), is written by Pål Strandbakken, and is an account of the testing of the third generation concept in the NanoDiode project; the Norwegian pilot and five subsequent European events.

The third chapter in Part IV, *Participatory democracy: Hybrid Forums and Deliberative Processes as Methodological Tools* (Chapter 15) is written by Virginie Amilien, Barbara Tocco, and Pål Strandbakken, with eight other European contributors from Strength2Food. In it, we introduce a version of Callon *et al.*'s 'Hybrid Forums', called HF 2.0, and compares it to the third generation deliberations.

The fourth, and final chapter in this part; *Conclusion 2020: A more Democratic Science Through Public Engagement?* is written by Pål Strandbakken, and looks at the relation between deliberative, or participatory democracy and the representative system, building on the experience from the analysed cases, after a comparison with HF 2.0 and a quick assessment of the relation to RRI.

PART I

SCIENCE AND DEMOCRACY

PART

SCIENCE AND DEMOCRACY

Chapter 1

Emerging Technologies, Deliberations, and Democracy

Pål Strandbakken

SIFO Consumption Research Norway, Oslo Metropolitan University,
Postboks 4, St. Olavs plass 0139 Oslo, Norway
pals@oslomet.no

1.1 Introduction

The events and processes that were analysed in the first edition of this book (Strandbakken, Scholl, & Stø, 2013) were selected because the editors believed that they would contribute to our understanding of the theme of scientific or technological democracy. In the present edition, we want to take the analysis further, also by building on the insights from more projects. Most important, we have had the opportunity to develop and test the so-called third generation deliberative process on nanotechnology, which was envisioned in the concluding chapter of the first edition (Chapter 13; Conclusions: Towards a Third Generation of Deliberative Processes). Further, the theme of democratic science and technology is broadened by the

Consumers and Nanotechnology: Deliberative Processes and Methodologies
Edited by Pål Strandbakken, Gerd Scholl, and Eivind Stø
Copyright © 2021 Jenny Stanford Publishing Pte. Ltd.
ISBN 978-981-4877-61-9 (Hardcover), 978-1-003-15985-8 (eBook)
www.jennystanford.com

analysis of a rather different method or approach used in this field, that is, the hybrid forum. We still restrict the analysis to countries usually described as democracies; that is, countries with universal suffrage, freedom of speech, and regular elections as a minimum requirement.

The idea that science is something that can (and should) be democratised might seem awkward or paradoxical for the following reasons:

1. The development of science and its subsequent transfer into technology is an extremely expensive and long-term activity, and it has mainly been the concern of intellectual or economic elites. It has rarely been a matter for broad political engagement.

2. To the extent that science is financed by parliaments on behalf of taxpayers, science in a way is under democratic control already...while privately financed research is seen as a part of the free enterprise system; hence, nothing the public (or the government) should tamper with, as long as the research activity observes laws and regulations.

This means that, at least at a rather abstract level, citizens already regulate science through their votes and finance it through their taxes. Parliaments and governments make rules, decide on priorities, and grant money on behalf of the voters. Basically, this is correct in principle, but it hardly makes science into something that citizens feel they control or influence. An additional constraint is that science and technology is not often politicised, in the sense that political parties debate alternative science and technology policies in their election campaigns. There has been some controversy, often voiced by political parties based on religious sentiments, over biotechnology and the new reproductive technologies, but most science development and most introduction of new technologies go unnoted or at least undebated.

The big exception is GMO (genetically modified organism). European citizen-consumers put up a strong resistance to it and tied the hands of the politicians. Some see this as a triumph for democracy and engagement, and others complain and regard it as an example of public irrationality; something that EU initiatives on RRI (responsible research and innovation) and science communication are designed to help us avoid in the future. In the present publication, we seek to introduce a more ambitious perspective on scientific democracy, ideally with a shorter distance between citizens on the one hand and policymakers and scientists on the other hand; one facilitating more nuanced debates on these complex matters, and one where positions are not immediately reduced to being *for* or *against* a given technology or a solution.

But questions of scientific democracy remain theoretically and politically rather awkward. On the one hand, science is supposed to be 'free', whatever that might mean. On the other hand, democratically elected governments and parliaments are funding scientific research and its infrastructure, such as educational systems and laboratories. This means that the 'republic of science' is politically controlled or contaminated already, at least to a certain degree. We believe that the most promising way of democratising science is to make science policy an integral part of the general political debate:

> Political parties should formulate their views and opinions on science, what their priorities are, and how these priorities will be reflected in budget proposals. Then we might politicise questions of total and relative expenditure; research vs. defence, health etc., and questions of what fields of research to prioritise, like solar vs. nuclear energy research etc.

The focus on politicising science and technology clearly makes emerging, hence 'unstable' (Akrich, 1992), technologies more interesting than the already established and standardised ones, such as the emerging field of nanoscience and nanotechnologies, as the positive visions and potential applications for nanotechnology apparently are without limits. This is the case for medicine and bio-nanotechnology and also for energy, information and communication technology (ICT), materials technology, and uses for the consumer industry; all will be heavily influenced by the nanotechnologies (Ratner & Ratner, 2003).

This review of the use of deliberative processes over emerging technologies, exemplified with nanoscience and nanotechnology, is focused on theories of democracy and on an idea of 'democratic science' (Cohen, 1989; Dryzek, 2002; Guttmann & Thompson, 2004). Beck's notion of *Risk Society* provides a kind of background or general perspective (Beck, 1992), with 'risk' being the shadow of 'unlimited potential'.

1.2 Studies of Science

The study of emerging technologies, if and how they stabilise, and how they interfere with existing technological and societal conditions has long been a part of *science and technology studies*, STS, (Bijker & Law, 1992), and especially STS in the actor–network theory (ANT) versions (Hess, 1997; Latour, 2007). Further, our concern with nanoscience and nanotechnology also touches on different aspects of public understanding and perception of science, aspects already covered within the STS tradition (Lewenstein, 1995; Wynne, 1995). There is no intention here of giving a broad presentation of the field of science in society studies and the range of different approaches. We mainly aim at giving a broad 'framing' of it and to highlight what we perceive to be a common underlying motivation behind many of these initiatives.

It seems as if theories and perspectives in these science, technology, and society fields are rather intermingled. Like when the technology assessment (TA) tradition advocates the use of deliberative processes modelled on the practice in the Danish Board of Technology under the heading 'Constructive TA' (Schot & Rip, 1997). One explanation for this might be that the most important impulse underlying all these approaches, traditions (and abbreviations...) really is shared:

> To reduce the human cost of trial and error learning in society's handling of new technologies, and to do so by anticipating potential impacts and feeding these impacts back into decision making and into actors' strategies (Schot & Rip, 1997, p. 251).

Modern takes on these themes will be found in the ELSI tradition – ethical, legal, and social impacts of technology – later usually renamed ELSA, substituting impacts with *aspects* (Rip, 2009). Since 2010, the key term is RRI, *responsible research and innovation*, a term that appeared in 2008 and has been common since 2011 (Tancoigne, Randles, & Joly 2014).

RRI has been defined as:

> (...) a transparent, interactive process by which societal actors and innovators become mutually responsive to each other with a view on ethical acceptability, sustainability and societal desirability of the

innovation process and its marketable products (in order to allow a proper embedding of scientific and technological advances) in our society (von Schomberg, 2011, p. 9)

According to Arie Rip, it still is not necessarily clear *what RRI is* (Rip, 2016), but it has nevertheless become a common point of reference, not the least in the EU program Horizon 2020.

To achieve better alignment of research and innovation with societal needs, a number of initiatives have been undertaken by EU Member States and the European Commission. These initiatives have shown that there is a need for a comprehensive approach to achieve such an improved alignment. Responsible Research and Innovation (RRI) refers to the comprehensive approach of proceeding in research and innovation in ways that allow all stakeholders that are involved in the processes of research and innovation at an early stage (A) to obtain relevant knowledge on the consequences of the outcomes of their actions and on the range of options open to them and (B) to effectively evaluate both outcomes and options in terms of societal needs and moral values and (C) to use these considerations (under A and B) as functional requirements for design and development of new research, products and services. The RRI approach has to be a key part of the research and innovation process and should be established as a collective, inclusive and system-wide approach (EU Report of the Expert Group, 2013).

Even with these, rather wordy explanations and definitions, it seems as if the core concern remains: how do we democratise science, how do we involve civil society, as stakeholders or as citizens in it, and how do we make science and technology more responsible and more responsive to societal needs?

1.3 Nanotechnology in Society

According to the visions, nanotechnology, as an enabling technology, will have an innovative influence on the production processes, energy and material use, and information and communication systems and, after a while, a substantial influence on the everyday life of individual consumers and households. 'In fact, one would be quite hard pressed to find a field which nanotechnology will not influence' (Ozin & Arsenault, 2005, p. viii).

We are promised cheaper, stronger, and lighter products, which means that in contrast to the previous history of technology, nanotechnology might actually help us combine economic growth with a reduced consumption of resources (Ratner & Ratner, 2003), perhaps most significantly energy. This apparent limitlessness of nanotechnology and science makes it hard to grasp, and it is not easy to directly compare it to other technological changes. But there is something to be gained from comparing it to one rather recent scientific revolution. The nanotechnology community can learn from the history of GMOs in Europe. The principal difficulty in delivering insights from biotechnology for nanotechnology is not their substantive differences, but rather a certain grand commonality: both are broad, diverse families of technologies. This is not like comparing the first telephone to the first telegraph. It is more like comparing a wide range of applications powered by electricity with a wide range of earlier applications powered by hydraulic, wind, and steam power. How do we summarise so many techniques and applications bundled under one name? How do we know which specific form of biotechnology is relevant to a specific form of nanotechnology (David & Thompson, 2008)?

We have observed a growing public scepticism to nanotechnology along two dimensions. The first is linked to a knowledge deficit concerning environmental hazards and possible health risks of the new nanotechnology materials. Second, nanotechnology raises fundamental questions on the relationship between man and nature (operating on the molecular level is perceived as 'tampering with creation'), and ethical, political, and even religious dilemmas are put on the public agenda. Both dimensions are present in recent discussions over actual and potential applications of nanoscale technology.

The political potential of public concerns over a new technology can be seen in the debate about genetically modified (GM) crops, a debate which some observers consider displaying interesting parallels to the emerging nanotechnology debate (Burke, 2004). The European consumers' resistance to GM crops and food has had serious consequences for plant research as well as for commercial development of new crops. Several developing countries in Africa refused to grow GM crops, chiefly for fear of being unable to export them to the European market (*The Economist*, 2002). As far as nanotechnology is concerned, some NGOs are already taking action, but few have called for a global moratorium. On the contrary, most stakeholders seem eager to avoid a new GM confrontation (Throne-Holst & Stø, 2007).

According to an old national survey undertaken in the United States in 2006, 42% of the population then had heard nothing about nanotechnology. In the meantime, the actual numbers will have changed, but the observation of how the public reacts to this level of information probably holds: The number of sceptics *increased* dramatically after respondents were exposed to just a little information (Peter & Hart Research Associates, 2006). This effect is confirmed in another survey from the United States; it appears as if learning more about nanotechnology makes people less favourable to it. The authors of that specific study explain this in terms of peoples' values (Kahan *et al.*, 2007); others regard the conclusion as being too simplistic (Cormick, 2009). In the regular societal debate over nano matters, the Australian section of *Friends of the Earth* has taken the lead, voicing a very critical stand towards nanotechnology, with a first focus on nanomaterials in sunscreens and cosmetics, but later expanding it to food and agriculture and to nanotechnology in general. Other NGOs and institutes have adopted similar views.

Consumers, citizens, and their organisations could be the most important stakeholders in the diffusion process of nanoproducts. Failure or scandals within one branch, or even for one product, might affect all branches dealing with nanotechnology. This is a possible scenario if we do not consider the ethical challenges among consumers from the beginning. On the other hand, the 'scandal' with Magic Nano did apparently not have such dramatic effects: The aerosol-driven oven cleaner 'Magic Nano' had to be withdrawn from the market because consumers reported respiratory problems after using it. Most likely, the product had nothing to do with the nanoscale, and the problems were caused by the propellant gas and not by the active substance. It is mentioned here because the health panic over a product carrying the nano tag did not lead to a general nano panic.

Nanotechnology and the Public Debate

Strangely enough, it is actually rather hard to pin down precisely *what* the public should debate, as nanoscience, nanotechnology, and nano-enabled products cover a very wide and heterogeneous field. This again has much to do with the choice of *when* one should introduce public deliberations; do we favour upstream, midstream, or downstream engagement (Delgado *et al.*, 2011). The initiatives analysed in this book appear to be potentially most influential when engagement is midstream, but 'upstream' debates on science policy appear to be more common.

One take on this might be the increasing specialisation of nano deliberations, the 'generations' that we identify in the old concluding chapter, and the comparison of our approach with a method like hybrid forum. To decide on the desirability of different applications and to assess a diversified field of potential future benefits and risks are unfamiliar challenges.

The parallel to the GMO debate has repeatedly been mentioned, as a lot of policymakers, stakeholders, and business leaders have regarded this debate and its outcomes as dysfunctional and as a failure of the system. This is just one of many possible views, however, depending on which perspective the observer chose. But one might agree that the GMO debate became too polarised too quickly. It seemed as if, after a certain point in time, one either had to be against the technology or in favour of it. This is rather unfruitful for a debate that deals with a kind of 'nuanced uncertainty'.

Further, if nanotechnology is expected to have a much more profound impact over a lot more sectors of science and society than GMO, it will have to be a different sort of debate, or rather a different set of debates. The value of popular and democratic participation in science debates could be, and are contested, as business leaders and scientists often would prefer to work and develop the field without interference from amateurs. Here, this view is not an option. Public participation might indeed be difficult to incorporate, but difficulties should not be an argument against applying it.

Lack of nano knowledge in the population is not the only knowledge- or consciousness-based constraint to a successful debate, however. Scientists' knowledge of public desires and needs is limited as well. But a heightened general awareness of nanotechnology and a certain level of knowledge of the phenomenon are commonly supposed to be a prerequisite for initiating an interesting debate between lay citizens and between citizens and experts. Sadly, it seems as if the representation of nanotechnologies in fiction, in popular science, and in the media still mainly deals with micro machines and assemblers (Crichton, 2002; Drexler, 1986; Gibson, 1996), while actual nano presence in ordinary life today is more about carbon nanotubes in squash rackets, sunscreens, and antibacterial sports socks. This discrepancy makes it difficult to define a set of themes around which to organise a debate.

Ideally, we should probably try to debate products and applications that are *between* the products presently on the market and the ones in the science fiction literature, if we aim at influencing technologies and decisions. This seems to be an argument for 'midstream' (and 'upstream'?) engagement (Delgado *et al.*, 2011). Co-creation, briefly mentioned in Chapter 16, is supposed to work very close to real innovation processes.

The discrepancy mentioned above between actual concern over assemblers and the quite rare concern over properties of materials in the nanoscale may materialise into two different sets of discussions, focusing on futuristic applications in one and on the present technology in another.

These more general and contextual concerns frame the subject matter of this chapter, being a kind of theoretical and conceptual introduction to a review of a set of deliberative processes organised in Europe and in the United States. It seems fair to mention, however, that this is not a theory-driven book. It is organised around the actual events, the background for them, the methods of conducting them, and the results obtained from them. Our aim is to contribute to bridge the gap between scientific and political matters in relation to nanotechnology.

1.4 Engagement 'Market'

The realisation that public engagement, based either on stakeholders or on citizens, is the clue to smooth nano governance (and a prerequisite for European funding...) has opened up a huge market for engagement methods and approaches. In this book, we cover deliberations (as consensus conferences and citizens' juries) and a version of hybrid forums (HF 2.0), while co-creation as a tool in innovation processes under RRI is mentioned.

In addition, there are lots of approaches, building on those mentioned, plus on 'public advisory committees', 'civic dialogues', 'focus groups' and so on. All might be categorized after their purpose, their representativeness and their degree of transparency and more. Some of the methods are branded by consultancies offering to organise engagement events for businesses, authorities, and civil society organisations (CSOs). We have seen handbooks for stakeholder engagement, directed at CSOs or businesses trying to get into research and technological development processes, and we have seen guides for researchers who want to come in contact with society. For this publication, however, we limit our analysis to the two mentioned approaches, even if we comment on standardisation in Chapter 11. For deliberations and hybrid forums, we have hands-on experience that we want to make available for everybody.

1.5 Deliberative Democracy and Governance

As alternatives to, or even contributions to the completion of, representative democracy, participatory, deliberative, and direct democracy are mentioned. The terms deliberative and direct do not cover identical ideas, but they are grouped together here because their relation to the representative system seems rather similar.

In European political philosophy, Jean-Jacques Rousseau is a common starting point for discussions over participatory democracy (unless one wants to take off from Athens in antiquity). Two things should probably be voiced against Rousseau's plans for a just society: First, his political theory seems rather unsuited for governing larger political units; it appears to be most relevant for villages based on agriculture and craftsmanship where each man represents himself. Second, his idea of the general will (*volonte generale*) is unclear, even mystical. It might be used to justify both tyranny and deliberative democracy. One reason for this is that a 'common interest' can be hard to outline and define when and where there are conflicts of interest (rural vs. urban, worker vs. employer, women vs. men etc.). Our distinction between consensus conferences (Chapter 3) and citizen juries (Chapter 4) partly refers to this. Rousseau's thoughts have inspired political movements like anarchism as well, but the most relevant participatory democracy initiatives in the 20th Century were linked to industrial democracy and workers' participation in decision-making (Pateman, 1977). Participation and deliberation in matters of science and technology have to be framed differently and organised in other ways, however.

In sociology and political science, much attention in later years has been put on the concept of governance of emerging technologies.

One important argument for organising deliberative processes with citizens and with stakeholders is that the absence of a political debate and the absence of public concern make it fruitful, perhaps even necessary, to *simulate* these political processes. The common voice will have to be represented inside political laboratories until it materialises in real NGO activity and in real politics. To the extent that the *Friends of the Earth* in Australia is able to politicise the nano themes and raise public and stakeholder awareness, deliberative processes on nano will be less necessary there.

The concepts of deliberation and deliberative processes have emerged from theoretical work on deliberative democracy. Deliberative, or discursive, democracy was coined by Bessette (1980) in the book *Deliberative Democracy*. It can also be linked to the work of Habermas (1989) and his attempts to define an ideal model for public debate. Deliberation can produce new options and new solutions, and it has the potential to document the full scope of ambiguity associated with the problem, which is relevant for our focus on the strong link between discourses of environmental problems and of nanotechnology.

For the reviews presented here (in Parts II and III), we chose to use the criteria for ideal deliberations presented by Cohen (1989) in his article 'deliberation and democratic legitimacy'. According to him, we should observe four criteria for ideal deliberation:

1. It should be a *free* discourse: participants should regard themselves as bound solely by the results and preconditions of the deliberation process.
2. It should be *reasoned*: meaning that parties are required to state their reasons for proposals.
3. Participants in the deliberative process should be *equal*.
4. Deliberation ought to aim at rationally motivated *consensus*.

His criteria were easy to apply to a large set of different situations and events, and they were quite simple to use. This simplicity helped us to avoid getting lost in theoretical and metatheoretical subtleties and debates. In a way, we have seen Cohen's criteria for deliberation as a way of operationalising Habermas. To apply his criteria seemed easier than having each author to struggle with their own interpretation of Habermas' original texts. The criteria actually seem to fit rather well with Habermas' thinking on the ideal conditions for societal debates; however, and they are relevant for both public and stakeholder deliberations. It will, of course, be difficult to reach these goals and ideals in practice, but this does not affect their status as 'ideals'. In the context of this book, the application of Cohen's criteria also helped make the different events comparable and, in addition, each of them point to important defining aspects of deliberations. In the new empirical chapters (14 and 15), we still have his criteria in mind, but we do not use them to explicitly organise the texts.

After working with these ideals in a context of real deliberations, it was clear that Cohen's criteria are not above critique and discussion. On the contrary, they often directly point to the relevant questions that should be discussed; that is, it is worth considering the extent to which achieving consensus always is the most desirable aim for deliberative processes. Consensus is one possible aim, but another is to tolerate, perhaps even encourage disagreement and to be aware of and respect the different positions of the stakeholders or citizens involved in the process. This means acknowledging the value of conflict in deliberative processes and recognising the reasons for disagreement rather than necessarily finding grounds for agreement. The productive value of conflict is highlighted in the chapter on hybrid forums (Chapter 15) and in the chapter on the UK NanoJury (Chapter 4). As the term jury indicates, majority votes on conflict issues are part of the procedure here.

Consensus should be distinguished from compromise. A compromise is a product of bargaining and belongs more to the concept of new governance. In terms of the degree of formal institutionalisation, the concept of deliberative processes is, to some extent, used with regard to processes with relatively low levels of institutionalisation, such as citizens' panels, public forums, and consensus conferences. We would, however, also see the concept as applying to more institutionalised activities, such as formal hearings and advisory committees. This means that we include the European and international standardisation processes as important deliberative tools for nanotechnology discourse, processes that are particularly relevant for the expanding consumer market (Chapter 11). In standardisation work, we find strong elements of stakeholder deliberation combined with citizen involvement. In addition, the outcome of such processes has a large impact on the field. This makes them natural candidates for our analysis.

An additional issue is that while deliberative processes are usually regarded as a supplement to normal democratic processes, they could also be seen as a way of undemocratically bypassing legitimate representatives of the popular voice. An example of this might be if, despite the official views of consumer organisations, environmental organisations and political parties being known, a deliberative process excluding them is arranged as a way of capturing a more positive 'public' voice.

We would have preferred to leave pure research processes out of our analysis, but when it comes to subject matter and methods, it is not easy to point to the difference between a focus group research session and a deliberative process. In our opinion, it would mainly bear on the strategic positioning of the event: a deliberative process is supposed to have an impact to influence policymakers directly, while a focus group which is part of a research process at best has a potential indirect influence, through the reporting and translation of public voices by researchers.

A deliberation, more than a focus group, has to deal with the problem of unfulfilled expectations. Participants feel that they are engaged in important matters and are easily disappointed over any subsequent real action.

A common feature of deliberative processes and focus group interviews is that it is necessary to supply information to participants early in the event in order to achieve an interesting exchange of views and information. We can imagine that the quality and possible bias of this information will influence the results of the deliberation, so that the outcome of a deliberation could be predicted. The character of the supplied information would therefore be a relevant consideration in our evaluative criteria, and even without any conscious manipulation, there is a question of how orchestrated the process is. Power, resources, and knowledge are not evenly distributed, no matter how neutral and unbiased an organiser tries to be, and the way in which this plays out in practice will be relevant to any review process.

The question of who is represented in a deliberative process can also be seen as problematic. On this, we would like to emphasise that we distinguish between two main types of deliberations: *processes aiming at the representation of a public or 'common' voice* (Part II) and *processes involving stakeholders,* who represent the interests of various 'constituencies' of business or political/organisational life (Part III and Chapter 14).

Classical representative democratic theory builds upon the ideal of one man/one vote and is based on the power of the majority. New governance and stakeholder approaches and deliberative processes offer supplements to traditional processes by acknowledging and giving legitimacy to lobbying, negotiations, and consensus-driven cooperation. A perceived shift from government to governance and to a new regulatory state presents an important development in legislation, regulation, and public policy in Europe (Majone, 1996, 1999).

The main idea behind the concept of governance is to involve stakeholders in sharing responsibility for political, economic, and judicial developments in societies, in dialogue with political authorities. In a White Paper on European Governance for the EU (COM, 2001), the main principles of governance were defined as *openness, participation, accountability, effectiveness, and coherence.* Discussions have also included notions of *democratic legitimacy* and *subsidiary* as further important principles. What roles do stakeholders have to play in the regulation of modern nanotechnology? Is it possible to identify these main principles of governance in current discourse on nanotechnology?

In the food sector, we have seen that industry and retailers have taken independent initiatives to develop standards and health-related schemes. These activities aimed at the enhancement of the consumer trust and increasing brand value in addition to avoiding litigation claims. These measures seem to coexist and partly overlap with public regulations in the same area (Marsden *et al.*, 2000). This kind of privatised regulation has been called self-regulation and is increasingly employed by the EU to regulate in a number of subject areas, for example, food safety and environmental standards (Majone, 1999).

Commentators claim that in some countries and some sectors, these private, often retailer-led, initiatives take on responsibilities public authorities would otherwise have to cover. In some countries, a pragmatic division of tasks and responsibilities seems to have evolved between regulating authorities and big business, saving public finances and maintaining markets for the companies involved. In this sense, co-regulation (Black, 2002; COM, 2001) and private interest regulation (Marsden *et al.*, 2000) have been suggested as appropriate terms for this situation. One should perhaps not be too naïve concerning the nature of a process in which business is kind enough to regulate itself, but if, on the other hand, the government decides on goals and ambitions, self-regulation might be cost-effective.

What do we mean by a *stakeholder* approach? The 'classical' concept of the stakeholder was developed within management theory on the relationship between business, on the one hand, and their environments, on the other hand. It was an expansion of the well-known *shareholder* concept: firms have to take into account not only the interests of their shareholders but also their stakeholders. In his book *Strategic Management: A Stakeholder Approach*, Freeman defines stakeholders as 'any group or individual who can affect or is affected by the achievement of the firm's objectives' (Freeman, 1984). During the last 20 years, this concept has been developed in various directions, and at least three of which are relevant here:

- First, we have seen the development of 'corporate social responsibility' (Carroll, 1999; Windsor, 2001), exemplified in the notion that businesses have responsibilities beyond their economic performance and should take into account other interests than those of their shareholders.
- Second, the concept has expanded from business management theory to society. It now integrates the responsibilities of organisations, policymakers, science, and consumers (Dentchev & Heene, 2003). This expansion has been controversial, but not without success.
- Third, we have seen a debate over the categorisation of various groups of stakeholders. The most relevant distinction is between primary and secondary stakeholders: 'A primary stakeholder group is one without whose continuing participation the corporation cannot survive. Secondary stakeholder groups are defined as those who influence or affect, or are affected by, the corporation' (Clarkson, 1995).

Deliberative processes, the stakeholder approach, and new governance all represent an alternative or supplement to representative democracy. While representative democracy has to be constituted around voting behaviour and the relationship between voters and their representatives, these models build their legitimacy 'on the degree to which those affected by it have been included in the decision-making processes and have the opportunity to influence the outcomes' (Young, 2000). The importance of this has been acknowledged in the representative democracies through the use of public hearings and the acceptance of organised lobbyists, but the other models take the involvement of stakeholders further.

An account of deliberative democracy grounded in communicative action in the public sphere requires some means of transmission of public opinion to the state. (----)... rhetoric is an important mode of communication here because it entails communication that attempts to reach those subscribing to a different frame of reference, or discourse. Thus, rhetoric plays an important role in deliberating across difference, as well as across the boundary between the state and the public sphere (Dryzek, 2000, p. 167).

Instead of waiting for the public sphere to form opinions and ideas on nano governance, the deliberations create artificial public spheres to simulate the processes in order to provide input to the political decisions that the legitimate representative democratically founded institutions have to regulate and decide on.

References

Akrich M (1992) The de-scription of technical objects, in *Shaping Technology/Building Society: Studies in Socio-technical Change* (Bijker WE, Law J, eds), MIT Press, Cambridge, Massachusetts.

Beck U (1992) *Risk Society. Towards a New Modernity*, Sage Publications, London.

Bessette JM (1980) Deliberative Democracy: The Majority Principle in Republican Government, in *How Democratic is the Constitution?* AEI Press, Washington, DC, pp 102–116.

Black J (2002) Critical reflections on regulation (Discussion Paper No. 4), London: Centre for Analysis of Risk and Regulation, London School of Economics and Political Science.

Burke D (2004) GM food and crops: What went wrong in the UK? Many of the public's concerns have little to do with science, *EMBO Report*, 5(5), 432–436.

Carroll AB (1999) Corporate social responsibility. Evolution of a definitional construct, *Business Society*, 38(3), 268–295.

Crichton M (2002) *Prey*, Harper Collins, London.

Clarkson M (1995) A stakeholder framework for analyzing and evaluating corporate social performance, *The Academy of Management Review*, 20(1), 92–117.

Cohen J (1989) Deliberation and democratic legitimacy, in *The Good Polity* (Hamlin A, Pettit P, eds), Blackwell, Oxford, pp 17–34.

COM (2001) European Governance. A White Paper. Commission of the European Communities 25.07 2001.

Cormick C (2009) Why do we need to know what the public thinks about, *NanoEthics*, 3(2), 167–173.

David K, Thompson PB (eds) (2008) *What Can Nanotechnology Learn from Biotechnology? Social and Ethical Lessons for Nanoscience from Debate over Agrifood Biotechnology and GMOs*, Food, Science and Technology, International Series, Academic Press, San Diego, CA.

Delgado A, Kjølberg KL, Wickson F (2011) Public engagement coming of age: From theory to practice in STS encounters with nanotechnology, *Public Understanding of Science*, doi:10.1177/0963662510363054.

Dentchev N, Heene A (2003) Toward stakeholder responsibility and stakeholder motivation: Systemic and holistic perspectives on corporate responsibility, in *Stakeholders, the Environment and Society: New Perspectives in Research on Corporate Responsibility* (Sharma S, Starik M, eds), Edward Elgar Publications, Northampton, pp 117–139.

Drexler KE (1986) *Engines of Creation: The Coming Era of Nanotechnology*, Anchor Books, New York.

Dryzek JS (2002) *Deliberative Democracy and Beyond. Liberals, Critics, Contestations*, Oxford University Press, Oxford.

EU (2013) *Options for Strengthening Responsible Research and Innovation. Report of the Expert Group on the State of Art in Europe on Responsible Research and Innovation*, Directorate-General for Research and Innovation Science in Society 2013, EUR25766 EN.

Freeman RE (1984) *Strategic Management: A Stakeholder Approach*, Cambridge University Press, Cambridge.

Gibson W (1996) *Idoru*, Viking Press, New York.

Guttman A, Thompson D (2004) *Why Deliberate Democracy?* Princeton University Press, Princeton & Oxford.

Habermas J (1989) *The Structural Transformation of the Public Sphere: An Inquiry into a Category of Bourgeois Society*, MIT Press, Cambridge, MA.

Hess DJ (1997) *Science Studies. An Advanced Introduction*, New York University Press, New York.

Kahan DM, Slovic P, Braman D, Gastil J, Cohen GL (2007) *Affect, values, and nanotechnology risk perceptions: An experimental investigation.* Available at https://papers.ssrn.com/sol3/papers.cfm?abstract_id=968652.

Latour B (2007) *Reassembling the Social: An Introduction to Actor-Network-Theory*, Oxford University Press, New York.

Lewenstein B (1995) Science and the media, in *Handbook of Science and Technology Studies,* revised edition (Jasanoff S, Markle G, Petersen J, Pinch T, eds), SAGE Publications, Thousand Oaks, CA, pp 343–361.

Majone G (1996) *Regulating Europe*, Routledge, London.

Majone G (1999) The regulatory state and its legitimacy problems, *West European Politics*, 22(1), 1–24.

Marsden T, Banks J, Bristow G (2000) Food supply chain approaches: Exploring their role in rural development, *Sociologia Ruralis*, 40(4), 424–438.

Ozin GA, Arsenault AC (2005) *The Nanotech Pioneers: Where Are They Taking Us*, The Royal Society of Chemistry, Cambridge.

Pateman, C (1977) *Participation and Democratic Theory*, Cambridge University Press, Cambridge.

Peter D, Hart Research Associates (2006) *Public awareness of nano grows – Majority remain unaware.* Woodrow Wilson International Center for Scholars, Center Project on Emerging Nanotechnologies, Washington, DC.

Ratner MA, Ratner D (2003) *Nanotechnology: A Gentle Introduction to the Next Big Idea*, Prentice Hall Professional, Upper Saddle River, NJ, USA.

Rip, A (2009) *Futures of ELSA: Science & Society Series on Convergence Research*, EMBO Reports 10(7), 666–670, Available at https://doi.org/10.1038/embor.2009.149.

Rip, A (2016) The clothes of the emperor: an essay on RRI in and around Brussels, *Journal of Responsible Innovation*, 3(3), 290–304.

Schot J, Rip A (1997) The past and future of constructive technology assessment, *Technological Forecasting and Social Change*, 54, 251–268.

Strandbakken P, Scholl G, Stø E (2013) *Consumers and Nanotechnology: Deliberative Processes and Methodologies*, Jenny Stanford Publishing (formerly Pan Stanford Publishing), Singapore.

The Economist (2002) Better dead than GM-fed? March 13, p 31. Available at http://www.economist.com/node/1337197. Accessed November 22, 2012.

Tancoigne E, Randles S, Joly P-B (2016) Evolution of a concept: a scientometric analysis of RRI, in *Navigating Towards Shared Responsibility in Research and Innovation. Approach, Process and Results of the Res-AGorA Project.* 39-46. ResAGorA.

Throne-Holst H, Stø E (2007) Føre var-prinsippet innen nanoteknologien: Hvem skal være føre var? Final report from the Nano governance project. Project report no. 7. National Institute for Consumer Research, Oslo.

von Schomberg, R (2011) Prospects for Technology Assessment in a Framework of Responsible Research and Innovation, in *Technikfolgen Abschätzen Lehren: Bildungspotensiale* (Dusseldorf, Beecroft, eds.), Wiesbaden: Transdisiplinärer Verlag Methoden.

Windsor D (2001) The future of corporate social responsibility, *International Journal of Organizational Analysis*, 9(3), 225-256.

Woodrow Wilson International Center for Scholars, Center Project on Emerging Nanotechnologies, One Woodrow Wilson Plaza, 1300 Pennsylvania Ave., NW, Washington, DC 20004-3027.

Wynne B (1995) Public understanding for science, in *Handbook of Science and Technology Studies,* revised edition (Jasanoff S, Markle G, Petersen J, Pinch T, eds), SAGE Publications, Thousand Oaks, CA, pp 361-389.

Young IM (2000) *Inclusion and Democracy*, Oxford University Press, USA.

Chapter 2

Overview of a Set of Deliberative Processes on Nano

Gerd Scholl and Ulrich Petschow

Institute for Ecological Economy Research (IÖW), Potsdamer Str. 105, 10785 Berlin, Germany

gerd.scholl@ioew.de

2.1 Introduction

In this chapter we account for the first batch of deliberative processes collected for the Nanoplat project. As a first step, an inventory of deliberative processes on nanoscience and nanotechnology was prepared. In its latest version this inventory contains more than 60 different processes conducted in Europe and abroad (mainly the United States).[1]

Based on this overview, more than a dozen of deliberative processes were selected for review, the selection criteria being such as broad European coverage, variety in types of deliberation, consumer goods focus, availability of data and access to the field. The final selection covers four processes at European and nine processes

---

[1]The inventory is available at www.nanoplat.org.

Consumers and Nanotechnology: Deliberative Processes and Methodologies
Edited by Pål Strandbakken, Gerd Scholl, and Eivind Stø
Copyright © 2021 Jenny Stanford Publishing Pte. Ltd.
ISBN 978-981-4877-61-9 (Hardcover), 978-1-003-15985-8 (eBook)
www.jennystanford.com

at national level. Most of the processes in this selection later made it into this book:

- Nanobio-RAISE (EU)
- Nanologue project (EU), Chapter 10
- Code of Conduct for Responsible Nanosciences and Nanotechnologies Research (EU)
- Standardisation (EU), Chapter 11
- PubliFocus (CH)
- Consumer Conference on Nanotechnology in Foods, Cosmetics and Textiles (DE), Chapter 5
- Citizens Nano Conference (DK), Chapter 3
- Citizens' Conference on the Nanotechnologies, Ile de France (FR), Chapter 7
- Conferences Cycle on Nanotechnology 'Nanomonde' (FR), Chapter 6
- Public Consultation on Nanotechnology for Healthcare (UK)
- Nanodialogues (UK)[2]
- Nanotechnology Citizens' Conference (US), Chapter 8
- National Citizens' Technology Forum (US), Chapter 9
- Website of the Meridian Institute (US)[3]

The point of departure for the review of deliberative processes on nanoscience and nanotechnology is the four criteria of Cohen (1989) on ideal deliberation:

- It is a *free* discourse; participants regard themselves as bound solely by the results and preconditions of the deliberation process.
- It is *reasoned*; parties are required to state reasons for proposals.
- Participants in the deliberative process are *equal*.
- Deliberation aims at rational, motivated *consensus*.

These basic criteria have been reflected against the results of current research on public engagement (Gavelin *et al.*, 2007;

---

[2]The NanoJury (Chapter 4) was part of the same broader nano initiative in the United Kingdom as the Nanodialogues.

[3]In this review different deliberation-oriented activities of the Meridian Institute were reviewed.

Warburton *et al.*, 2007, 2008) and further developed into a list of guiding questions for performing the reviews. Besides a section of the review in which the deliberative process under consideration is depicted in its basic characteristics (e.g., initiator, funding, purpose, timeline, main topics, etc.), each review addresses the following issues:

- *Initiation*, for example, At what point of policy decision-making is it located? In what way is the process intended to give advice on how to proceed further?
- *Organisation*, for example, What are the monetary costs of the process? Is the process equipped with a reasonable amount of resources? Is the scale of the process appropriate to its purpose and objectives?
- *Content*, for example, Are the scope and the content of the process clearly defined and explicitly declared to the participants? Are there any thematic restrictions, and if so, how are they explained or legitimised?
- *Participation*, for example, Who are the participants in the process? How are they selected and by whom? What is the role of the different types of participants? How are the different participants practically involved in the process?
- *Reasoned process*, for example, What forms of interaction between participants are employed in the process? How are scientific and other reasoned arguments brought into the process? Is there a variety of opinions included? Is there room for mutual learning and change in positions?
- *Results*, for example, What are the outcomes of the process? Does it resemble a 'common voice' or does it reflect articulated interests of stakeholders? Who are the addressees of the outcome? Is there a movement towards some sort of shared opinion?

The reviews were based on secondary analysis (e.g., analysis of documents and evaluations if available) and a limited number of interviews with people involved.

In the following sections, we present the insights gained from the reviews along the above list of review criteria.

2.2   Major Findings from Reviews

2.2.1   Initiation

There is a *wide spectrum of organisations driving public engagement on nanotechnologies*, such as academia (universities, research institutions, etc.), policy consultants and policy advising research bodies (e.g., Danish Technology Board), professional engagement facilitators (e.g., Vivagora), public authorities (e.g., German BfR), research councils (e.g., UK EPSRC), etc. The initiators show varying scope in their decision-making — from informing the general public and/or stakeholders to funding research — which, of course, influences the potential *impacts* of the deliberative process.

There are different *purposes* on which deliberative processes are enacted. It can be about a general identification and assessment of public attitudes towards a certain technology, about experimenting with a new form of public dialogue in order to learn about its potentials and shortcomings, about informing a specific decision, for example, on research funding, from a citizens' perspective, etc.

In some cases the idea of *experimentation* with novel forms of public engagement was important. Hence, the question of how the process can be organised in an appropriate fashion came into the focus. This reveals that public participation and deliberate processes actually do not follow a given format. Instead one observes a search for adequate processes for specific purposes (e.g., public participation as an input to the development of fundamental research programs).

The clearer the (stated) *goals* of a deliberative process and, hence, the role of the stakeholders/citizens involved, the bigger its potential benefit. A lack of clear goal definition may lead to lack of motivation and, hence, impair participation. This holds true for initiators, organisers as well as for participants.

2.2.2   Organisation

The organisation of a deliberative process is a *costly exercise* (e.g., National Citizens' Technology Forum: $US500,000–600,000, German Consumer Conference: €100,000). Particularly, when taking into account personal involvement of staff from the initiating organisation (see, e.g., UK EPSRC case). Accordingly, one may assume that a reasonable cost-benefit-relation will strongly influence the further proliferation of public engagements. However, this goes along with the challenge to clearly assess the benefits of deliberation.

Rather important for the success of the deliberate processes are the *organisers and facilitators*, especially with respect to the recruitment of participants as well as to keep the participants motivated during a series of events. It seems important to have professional organisers which ideally have experience in nanotechnology, too.

The reviews made clear that different *forms* of deliberate processes are used: from a two-hour card game on nanotechnologies, one evening event (mostly information and discussion with experts) and three-hour long focus group discussions to processes running over half a year with three weekends (face-to-face) and interaction between meetings. Furthermore, the methods and tools used in the exercises also differ (see below).

2.2.3 Content

Consumer products are not very often at the focus of the deliberative processes reviewed. One may observe that this slightly changes during recent years, that is, with increasing market penetration of nano-enhanced consumer goods.

As mentioned above, if *goals and objectives* of the deliberative process are not clear and not clearly stated to the participants, the process runs the risks of producing poor results. However, the formulation of clear objectives is more difficult when the process set-up is more experimental.

2.2.4 Participation

Citizens did not take part in all deliberative processes reviewed. In case they did, they were mostly recruited in a way to achieve a wide demographic coverage. In some processes highly different segments of society took part. In case of citizens' involvement their roles were usually properly defined. Questions, such as the following, arise in this context:

- Is there a demographic bias of laymen towards better educated ones? How can demographic representativeness be ensured by recruitment?

- How important is financial compensation for ensuring commitment? (See National Citizens' Technology Forum: two weekends, nine two-hour online sessions, compensation for time and effort $US500!)

Experts were recruited in a way to achieve wide coverage in disciplines and expertise. Their roles in the process were usually properly defined.

Participation of the initiating organisation in the process provides the possible benefit of institutional learning (e.g., United Kingdom: Public Consultation on Nanotechnologies for Healthcare). There are other cases, however, where initiators intentionally refrained from participation in order not to influence the discussions between the laymen.

2.2.5 Reasoned Process

There is a *variety of tools employed to stimulate interaction* between participants, such as working groups, public hearings, plenary discussions, presentation plus questions and answer session, scenario techniques, card games, etc.

Scientific evidence is usually provided by different experts covering diverse backgrounds (interdisciplinarity). The laymen are not involved in or in charge of selecting the experts in all cases.

A *variety of opinions* is ensured by involving experts with different backgrounds and a diversity of laymen. The room for mutual learning and change in positions, however, strongly correlates with the *duration of actual deliberation* — which varies significantly, from three-hour evening sessions to processes comprising three weekend sessions and time for individual deliberation in between the events.

Apparently, the deliberative processes reviewed achieved on average a *good level of transparency* by means of, for example, proper documentation, clear communication to participants, dedicated press work, ex-post evaluations, etc. Organisers/initiators seem to be aware of the importance of transparency for a proper and effective deliberation. However, there is still room for improvement.

2.2.6 Results

The results of the deliberative processes reviewed are numerous: the *direct and tangible* ones encompass votes, recommendations, reports, etc. The *indirect and intangibles* ones are learning of participants, including awareness and sensitivity with respect to chances and risks of nanotechnologies, learning on how to manage and employ deliberative processes, trust building into public risk assessment and management, etc.

The *actual impact(s)* of the deliberations reviewed were difficult to assess due to a lack of data, lack of goal specification, lack of (information about) dissemination activities, etc.

Institutional learning is apparently one important impact for (semi-) public initiators. In a corporate context, however, this seems to be less easily achievable.

If *policymakers* are not or only loosely linked to the deliberative process, the actual impact on (their) decision-making is obviously very small. This appears to be the case particularly for deliberative processes driven by academia (e.g., U.S. National Citizens' Technology Forum).

A *prerequisite* for a significant impact would be a description of a clear avenue on how the deliberative process is going to influence policymaking; often we encountered a lack thereof. Furthermore, the participants in the process are usually not informed about 'the challenges facing the transmission and translation of their recommendations in actual practice' (Wickson, 2008).

With regard to the results provided by the deliberations there are hints that a poor follow-up might induce a *backlash*. Wickson (2008) reports that 'after this deliberative process (National Citizens' Technology Forum) the participants reported a decrease in their belief that their opinions or actions can actually affect political outcomes'. This raises the question of how to deal with the legitimate expectations of participants that their efforts have some kind of impact?

There is only little evidence provided by the reviews with regard to how *consensus* is being dealt with. It appears that if a 'common voice' of, for example, citizens-consumers, is sought of, a consensus among the group is also aspired to — normally including 'sensitivity to digressions from consensus' (*ibid.*).[4]

---

[4]For example, according to one report on the National Citizens' Technology Forum in the United States, 'For the purposes of this document, consensus indicates not unanimous support, but the wisdom of the group without major objection', see NCTF Wisconsin (2008).

2.3 General Conclusions

Lay or public involvement in advanced science and technology policy is practiced and actually possible. Its benefits are the biggest in issues of strategic relevance, that is, topics that are inherently interdisciplinary and have clear societal implications.

Figure 2.1 illustrates that deliberative processes in our sample differ in terms of length of the process and people involved (cp. Warburton *et al.*, 2008, p. 4).

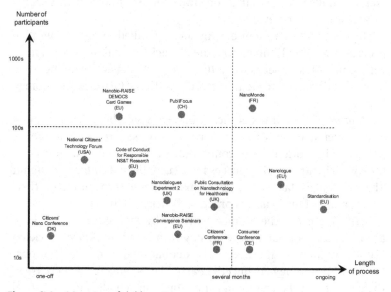

Figure 2.1 Mapping of deliberative processes.

There seems to be a *trade-off* between intensity of deliberation and how generalisable were its results. The less the people involved, the more intensive the interaction and deliberation can be. But the less the people involved, the smaller is the representativeness of the results (e.g., recommendations) (see, e.g., PubliFocus in Switzerland, cf. Rey, 2006, as opposed to Consumer Conference in Germany, cf. Chapter 5).

Clear avenues and standards concerning the further consideration of *results* of deliberate processes should be developed. This is sometimes the case, if, for example, public authorities initiate processes or if there are strong linkages to the parliament (Danish case). In other cases, there are no direct links to any political or administrative authority and, hence, deliberative processes have no clear addressee and might mainly be seen as one of many inputs in public discourse on nanoscience and nanotechnology.

The analysed deliberative processes sometimes have the main focus just to identify possible conflicts in the development of nanotechnology (avoid the GM 'faults'), while other deliberative processes are more interested in some form of more 'democratic' or democratically informed technology shaping. This might be at odds with democratic institutions and institutional arrangements (stakeholder inclusion).

The reviews reveal that deliberate processes might have a direct impact on decision-making, and they provide a few hints as to the influence of the country-specific political culture on actual impacts. In a country such as Denmark or, in particular, Switzerland, the political culture strongly relies on the inclusion of the citizens, and the national parliaments are well prepared to take recommendations gained from deliberate processes into account. In other countries, for instance, Germany and the United States, this kind of participatory democratic culture seems to be less developed and less established. Accordingly, the results of public-engagement exercises are less easily fed into political decision-making. But we have to point out that these observations are tentative and based on little information in the reviews that did not explicitly address the contextual issue of political culture.

References

Cohen J (1989) Deliberation and democratic legitimacy, in *The Good Polity* (Hamlin A, Pettit P, eds), Basil Blackwell, Oxford, pp 17–34.

Gavelin K, Wilson, R, Doubleday R (2007) Democratic technologies? The final report of the Nanotechnology Engagement Group, Involve, London.

National Citizens' Technology Forum (NCTF) Wisconsin (2008) National Citizens' Technology Forum: Final Report from Madison, Wisconsin, available at http://www4.ncsu.edu/~pwhmds/California%20 Final%20Report.pdf, last accessed 21.11.2008.

Rey, L. (2006) Public Reactions to Nanotechnology in Switzerland. The findings of the publifocus discussion forum 'Nanotechnology, Health and the Environment', Centre for Technology Assessment, Bern, available at http://www.ta-swiss.ch/incms_files/filebrowser/2006_ TAP8_Nanotechnologien_e.pdf, last accessed 16.11.2012.

Warburton D, *et al.* (2008) Deliberative public engagement: nine principles, Background Paper, N. C. C. Involve, London.

Warburton D, *et al.* (2007). *Making a Difference: A Guide to Evaluating Public Participation in Central Government*, Involve, London.

Wickson F (2008). *Review of the National Citizens' Technology Forum, Bergen*. Unpublished raw data.

PART II
CITIZEN-ORIENTED DELIBERATIVE PROCESSES

Introduction to Part II

Pål Strandbakken
SIFO Consumption Research Norway, Oslo Metropolitan University,
Postboks 4, St. Olavs plass 0139 Oslo, Norway
pals@oslomet.no

As indicated in Chapter 1, we find it necessary to distinguish between the kinds of deliberations where the organiser wants to consider the opinions and reactions of lay citizens, from the processes where the idea is to include the knowledge of, and to consider the interests of stakeholders. Organisers and participants in both types of events would probably state that their intention is to develop democracy, but citizens' influence and stakeholder influence should be regarded as two different approaches: the first is a layman approach, while the second is about incorporating organised interests, often with some professional background in the theme the deliberation covers. In empirical, as opposed to theoretical, deliberations, things are a bit less clear-cut, however. In most of the citizens' deliberations, stakeholders were invited to present their views and perspectives, but this was conceived as input to the debates and deliberations of the non-professionals.

We have also seen that the citizen-based deliberations vary along a number of dimensions (see Chapter 2); the number of participants, the time resources, participants' access to expertise, and the process' closeness to 'legitimate' policymaking. It is fair to mention that time resources cover both the duration of the deliberation itself, and the time available for mutual and individual learning (prior to the deliberation, between meetings). Much of this variation is present in this selection of seven deliberations. We have chosen to review the three 'classical' nano deliberations in Denmark, the United Kingdom, and Germany. This is also a result of increased intensity and increased access to resources; from the 13 persons/3 hours/2 lectures in Copenhagen via the U.K. NanoJury's 16 participants having 18 meetings with expert witnesses over a 5-week period to the German Consumer Conference's 16 consumer-citizens' 8-month engagement with access to a number of experts at different stages of the process.

The next two deliberations that are reviewed are the two French events/processes Nanomonde and the Citizens' Conference in Île-de-France. They are interesting because they are different in size, meaning, the number of participants, as well as on the question of intensity, that is, the duration of the process. In addition the relationship between the deliberation and the political authorities is very different. Beyond the scope of this publication, France is also an interesting case because of the significant nano unrest that has been reported after, but probably not because of these exercises in participatory democracy.

Finally, we review two processes undertaken outside of the European Union; a rather small and intensive process in Madison, United States, with 13 citizens meeting in 3 to 4 hour sessions, and the more ambitious National Citizens' Technology Forum (NCTF), also in the United States, where 6 groups of maximum 15 citizens in different geographical locations participated in 9 online sessions over a month-long period, in addition to face-to-face meetings in the groups. The NCTF is also notable for its rather precise theme of human enhancement. With the possible exception of the German Consumer Conference that concentrated on three important product groups, the other deliberations reviewed here approach the broad theme of 'nano and society'. Thus, the NCTF points towards what we regard as second (or even third) generation of deliberative processes (see Chapter 13). Because of the design of the forum, it is also the best documented case here, mainly because of a pre- and post-event questionnaire.

For Île-de-France and Madison, we have included the written statements from the groups as appendixes to the respective chapters, in order to illustrate the way that citizens formulated their political and ethical concerns and the scope of their engagement.

We have presented the events/deliberations in a more or less uniform way, but hopefully without strangling the authors' creativity too much. Where they felt it was needed, authors were free to use the introduction or the summarising appraisal to put the specific deliberation into political, national, and/or academic contexts. The most important common threads is that all chapters consider Cohen's four deliberation criteria, in addition to explaining matters of initiative, organisation, resources, and participation. Finally, we have registered whether consumer questions were explicitly treated in the deliberations.

Chapter 3

Citizens' Nano Conference in Denmark

Eivind Stø
National Institute for Consumer Research (SIFO), Postboks 4682 Nydalen, 0405 Oslo, Norway
eivind.sto@sifo.no

3.1 Introduction

The Danish deliberative process discussed in this chapter took place as early as June 2004. The event was organised by the Danish Board of Technology (DBT) as a one-day meeting with 29 citizens from the Copenhagen area. The applied methods were group discussions and questionnaires. In this chapter, we will discuss this specific consumer meeting. However, we will also consider the more general activity of the DBT related to nanotechnology and other emerging technologies like information and communication technologies (ICTs) and genetically modified organisms (GMOs). The chapter is based on the written report from the citizens' meeting, interviews with those responsible at DBT, a workshop arranged by the Board and the (Norwegian) National Institute for Consumer Research (SIFO) in November 2009 as well as some other reports from DBT.

Consumers and Nanotechnology: Deliberative Processes and Methodologies
Edited by Pål Strandbakken, Gerd Scholl, and Eivind Stø
Copyright © 2021 Jenny Stanford Publishing Pte. Ltd.
ISBN 978-981-4877-61-9 (Hardcover), 978-1-003-15985-8 (eBook)
www.jennystanford.com

After 2004 the DBT has remained visible in the public nano-discourse with a number of initiatives. In 2006, they invited 20 stakeholders to a workshop with external experts on nanotechnology in order to formulate a set of political conclusions and proposals; in 2007 it even used a theatre play to discuss nanotechnology. We will return to these activities later.

The method with citizens' involvement was developed already in 1987, and since then various changes and improvements have been introduced. In written statements from deliberative processes related to nanotechnology in other countries, we often find references to the *Danish Model* (Einsiedel *et al.*, 2001; Powell & Kleinman, 2008). The Danish model is basically the employment of deliberating citizens in consensus conferences, but it also refers to a more general inspiration from Danish authorities' attempts at involving citizens and arrange experiments on participatory democracy.

The board has played an important part in the development of the consensus conference model since the institution was established in 1985. However, it is perhaps worth noting that DBT itself refers to experiences from the United Kingdom and the United States (Klüver, 2009). But some of the roots of the deliberative processes may also be found in local Danish traditions (Andersen & Jæger, 1999; Klüver, 1995, 2009). To a certain degree, Denmark's position on citizens' involvement is embedded in the country's political and organisational culture, exemplified by the 'Folk high-schools' and 'citizen houses', established between 1850 and 1900. Some important contributions to this tradition are given by Grundtvig, a famous priest/theologian and philosopher. Citizens' involvement was also a natural part of the bottom-up political movement in the 1960s and 1970s (Andersen & Jæger, 1999).

The DBT has developed two main models for deliberative processes, *Consensus Conferences* and *Scenario Workshops* (Andersen & Jæger, 1999). In both models we find a combination of involvement from citizens and experts. One obvious difference between the models is suggested by the titles of the initiatives; another is that consensus conferences are public meetings while scenario workshops are closed events. The ideal model of a consensus conference includes a citizens' panel and an expert panel, run by an advisory or planning committee. It ends with a public meeting.

However, the actual conferences and panels may vary substantially because of variations in resources and deadlines.

The Danish Technology Board (DTB) was established in 1985 as 'Teknologi-Nævnet' and was reorganised in 1995 when it changed its name to the Danish Board of Technology (DBT). It is a self-governing independent institution connected to the Ministry of Science, Technology and Innovation in Denmark. It is also closely linked to the Danish Parliament through its Committee for Science and Technology. These two institutions have the responsibility to appoint the board members of the DBT. The Board received 13 million Danish Krones (approximately 2 million Euros) from the Parliament in 2008. According to its legislative mandate, the DBT's task is to organise independent technology assessments, carry out all-round assessments of the potentials and consequences of technology, initiate activities relating to public enlightenment education and communication, and advise the Danish Parliament and government.

The main purpose is to give critical advice to public authorities on problematic science and technological questions. Once a year the DBT reports to the Danish Parliament, Folketinget. It is worth noting that the topics vary substantially, from healthcare (2008) to transport (2005), from infertility (1993) to obesity (2009). DBT has also carried out hearings on alternative medicine and combating terrorism. Nanotechnology is only one among the large number of topics, and there is no permanent project on nanoscience or nanotechnology.

3.2 Description of the Process

The citizens' meeting on nanotechnology in Copenhagen in June 2004 was the first deliberative process in Denmark concerned with nanotechnology, and also perhaps the first processes ever, worldwide. The initiative was taken by the Danish Ministry of Research which also financed the process (Teknologirådet, 2004).

The main goals of this citizens' meeting on nanotechnology were not specified. The conference was not explicitly designed as a producer of input to actual decision-making processes; even though the Ministry of Research later used the report from the conference as input to a Parliamentary document on priorities for Danish research on nanoscience and nanotechnology.

The point of departure for this specific initiative was the view that nanoscience and nanotechnology were about to break through as important new scientific developments, while they were still unknown to most citizens. Further, the DBT emphasised that nanotechnology and nanoscience was a relatively broad concept that covered a variety of technologies. Finally, both the technology's positive potentials and its possible risks were to a large degree unknown to citizens, political authorities, and stakeholders at the time, and apparently also to the scientific community.

In the report from the citizens' meeting, the DBT refers to two rather diffuse arguments behind this initiative:

- By inviting ordinary citizens to this meeting the DBT wanted to broaden the political discussion because ordinary citizens may have other perspectives on nanotechnology than politicians, NGOs, experts, and scientists.
- At the end of the day it is ordinary citizens and consumers that have to live with the consequences of new technology.

It seems that the DBT wanted to use this meeting to stimulate a general political discussion on nanotechnology rather than using it as an input for specific decisions on regulation of nanotechnology. On the other hand, it certainly had an effect on the work and priority of DBT.

The time line of this process was relatively short. The meeting took place on the evening of June 7, 2004, from 17:00 to 20:00 hours. The participants only met this evening. This short meeting was organised in three steps: the first step was a plenary session where the participants were introduced to the topic through two lectures: one from the natural sciences and one from the social sciences. The first lecture dealt with the future potential of nanotechnology, the second with the softer ELSA[1]-related questions; here ethical, environmental, risk, and health aspects of nanotechnology were discussed (see Chapter 1).

In the second step, the 29 participants were divided into four groups who discussed the topic for an hour, before presenting their views at the plenary sessions, chaired by a moderator. In the last step the participants filled out a pre-coded questionnaire. This

---

[1]As mentioned in Chapter 1, ELSA is an abbreviation of ethical, legal, and social aspects of nanotechnology, under which the aspects that go beyond the natural sciences are considered.

questionnaire focused on knowledge of and attitudes towards nanotechnology. The main topics were health, environment and energy, the situation in the third world, the war and terror, and, at last, potential improvements of consumer goods.

The report from the DBT is based upon the discussions in these four interview groups and the 29 answers to the questionnaire.

Prior to the meeting, the participants had received some reading material on nanotechnology. This material was of 13 pages, with information suited for lay public, but it was meant for the participants to at least knew what kind of process they were taking part in and its theme. Participants were free to read and prepare themselves for the process, and some of them actually expanded their reading about nanotechnology beyond the material from DBT before the meeting.

The meeting was organised by the DBT. However, the *Danish Technology University* was involved in the process and so was the *2.0 LCA Consultancy*.

The topic of the meeting was general; it was not limited to specific areas or types of applications of technologies. The discussion concentrated on visions of nanotechnology on the one hand and ethical, social, and environmental aspects on the other. No specific consumer goods were discussed, but consumer products were a part of the discussions.

The aim of this process was the 'representation of a common voice'. The participants were not selected because they represented specific stakeholder's interests.

3.3 Review of the Process

3.3.1 Initiatives and Objectives

The initiative to citizens' nano conference was taken by the Danish Ministry of Research in dialogue with the DBT, and the ministry also financed the process. The aims of the activity were both advising the responsible ministry for research in Denmark and producing general input to the political discourse.

The DBT is responsible for running the scientific and political discourse about technology in general, and to nanotechnology more specifically, and it wanted to use data from this citizen's conference in the decision-making process about priorities for further dialogue with and among individuals on nanotechnology. However, the board is not responsible for the actual decision-making in this matter. This responsibility lies with the Danish Government and the Danish Parliament. The board functions as an independent advisory institution, and the conclusions from the meeting had potentials to influence further discourse on nanotechnology in Denmark.

This means that this specific consumer meeting on nanotechnology was not a single initiative to involve citizens in the discussion of emerging technology. It was part of a larger strategic program where the DBT was given the responsibility to expose ordinary citizens to relatively complicated technological questions. In two decades the DBT has generated trust among citizens, stakeholders, the scientific community, and politicians in science and technology questions, and it also has given legitimacy that they were responsible for running this first deliberation on nanotechnology.

However, the DBT is not running a permanent deliberation on nanotechnology. It has followed up with a classical stakeholders' conference in 2006 and an interactive theatre play on nanotechnology in 2007.

3.3.2 Organisation

One should remember that this nano conference took place in 2004, and this is reflected in the organisation of the process, in the questions put forward in the citizens' meeting, as well as in the answers from participants.

It was a relatively simple process, compared with other processes reviewed and discussed in this book. Administrative and financial resources were quite limited. However, this specific version of the Danish model has been adapted and used without large resources in other countries and in other situations. It is difficult to estimate the exact costs of the process, but as mentioned there are reasons to believe that the total economic costs were relatively small. This was confirmed by the board.

The meeting itself lasted for three hours and involved the staff of DBT and two external experts. Some resources were used for recruitment of participants and planning of the meeting. The meeting had to be documented, but this also seems to have been done with rather limited resources. If more resources had been available, the conference would probably have been redesigned.

Given the limitations, we still believe the resources were adequate. This was one of the first steps in involving individuals in the nanotechnology discourse and decision-making process, and the resources were sufficient to create necessary experiences. In many ways the Danish conference has functioned as an inspiration to other similar processes. In that sense the scale of the meeting seems appropriate to its purpose and objectives.

On the other hand it is possible to argue that three hours for such a meeting is insufficient to create fruitful responses from the participants. There is no real process, and the dialogue among the participants is limited within this design. In the first phase they listened to presentations from experts. In the second phase they participated in group interviews, and in the last phase they gave answers to a questionnaire. Only the second phase was a free discussion.

There were no thematic restrictions; it was a general discussion about advantages and challenges with nanotechnology, from a natural science perspective and an ELSA perspective. However, these general discussions were made more specific during the process; and this is certainly the case for the consumer questionnaire. It covers medicine, environment, energy, consumer products, and improvement of economic and social development in the third world. This reflects the emerging stakeholder and political discourse in 2004.

It is worth noting that consumer products were one of the topics in the questionnaire, even though very few products based on nanotechnology actually were found on the consumer market in Denmark at that time. So, the questions about consumption were hypothetical and more linked to the future possibilities than to actual products.

The scope of the discussion was formulated in three waves: (1) in the invitation to the conference, (2) in the 13 pages documentation participants received before the meeting, and (3) in the introduction to the conference and by the moderator in the group interviews. The content reflected the political discussion in 2004, and we believe that the participants knew what kind of process they were taking part in. This distinguishes deliberative processes from research-oriented focus groups where the participants only have a vague idea about the topic of the meeting. In this case it was possible for the 29 participants to prepare themselves for the consumer meeting, and some of them actually did so.

3.3.3 Participation

The 29 participants were ordinary citizens in the Copenhagen area, without any knowledge in nanoscience and nanotechnology, and it was an open meeting. The organiser seemed to be satisfied with the composition of the meeting, and no special initiatives were taken to include missing groups.

They were recruited in combinations of various methods:

- By advertisement in the free of charge MetroXpress, once (9 people)
- By using the DBTs homepage (1)
- By using the electronic newsletter from the DBT (2)
- By networking among the employees (10)
- By word of mouth (7)

The idea was that the participants should broadly reflect the Danish population with regard to gender, age, education, and religious affiliation. The conference was meant to be a meeting place for various voices, but of course, it is not meant to be a representative meeting. As seen below, along some variables the composition fits well with the Danish population, while others are more problematic:

- Among these 29 people 15 were women and 14 men
- Young people were slightly underrepresented in the meeting
- Participants had a significantly higher level of education than the majority
- Participants also seemed to be more oriented to the political left that the Danish average (but ten of the participant did not answer this last questions)

In addition it is worth noting that eight of the participants looked upon themselves as 'religious' people, most of them Christians. However, the religious dimension, which has proved important for how people approach nano, did not play an important part in the actual meeting.

It is our impression that the method used for selection of participants reflects the resources used in this deliberative process. Cheap methods, like networking among employees and word of mouth accounts for 17 of the 29 participants. However, this selection method seems sufficient to bring together people from different social backgrounds to discuss the emerging nanotechnology.

One of the important criteria for deliberative processes is that the process is reasoned. Participants are encouraged to argue for their views, not only to cast their votes. Knowledge then becomes a crucial factor for a successful process.

A pre-study showed that Danish citizens in 2004 knew little or close to nothing about nanotechnology (Teknologirådet, 2004). In an informal and semi-structured interview, 40 people were asked about their knowledge concerning nanotechnology. Half of them had heard about the concept of nanotechnology, but of them, only half knew that it had something to do with small scales. This pre-study was used as input to the design of the citizens' meeting. For a meaningful and reasoned process to occur, the participants had to increase their knowledge about nanoscience. This was a critical part of the process because the choice of information may influence the participants' attitudes.

At the citizens' meeting, the DBT had selected two introductory lectures to inform the participants. The speakers were not chosen in a dialogue between the citizens and the board, the responsibility was of DBT's alone. The first lecture was from the Danish Technological University and presented the theme from a natural science perspective. The second, a special consultant, presented the participants for the social, environmental, legal, and ethical challenges (ELSA perspectives). This was how the discussion within the scientific society was brought into the discussion.

This means that the citizens were not actually presented with different views, neither within the natural science nor the social science paradigm. These two one-dimensional information units may obviously have influenced the discussion in the group interviews following the presentations, but given the limited time available, this solution was probably not too bad. At least, the two speakers introduced the participants to different perspectives.

3.3.4 Results of the Process

The idea behind the citizens' meeting was to resemble a 'common voice'. However, this does not mean that this was a *consensus* conference. During the interviews the moderator did not search for any common agreement among the 29 participants. The design of the citizens' conference was actually the opposite. The DBT wanted to identify various perspectives and attitudes among the participants.

The outcome of the process was addressed to the board. The DBT on its side reported directly to the Danish Government and the Danish Parliament.

The results from this citizens' conference were reported to the Danish Ministry for Science and Technology. The ministry used the experiences and the report from this process very actively in their plan for research on nanotechnology and nanoscience in Denmark in December 2004 (Danish Ministry for Science and Technology, 2004). This strategic plan explicitly discusses ethical and environmental risks with reference to the involvement of citizens. Furthermore, these topics also became a part of the future research program in Denmark.

The more substantial conclusions from the citizens' meeting were results of the last two steps. It is no surprise that the majority of participants knew very little of nanotechnology and nanoscience when they volunteered for the citizens' conference. The group discussion started with a brainstorming session, and it showed an overwhelming general positive response to the development of nanotechnology. At the same time the citizens' conference asked for more research on risk assessment. In the report from the meeting it was also obvious that the participants were uncertain about the visions of nanosciences.

In one of the groups the participants wanted to spend 1/3 of the research money on ELSA aspects with nanotechnology. This was also supported by other groups, but not to the same degree. However, the main conclusion in 2004 was the uncertainty of this new technology.

This becomes even clearer when we look at the results from the questionnaire. The majority of participants share the idea that nanotechnology will contribute to (1) improve the conditions in the third world and (2) will contribute to reduce the potentials of war and terrorism. They also had positive visions for solving environmental problems and develop new energy resources. Furthermore, the potential for successful fights against various diseases and climate changes was also regarded as a strength of nanotechnology.

On the other hand, participants did not think about the importance of using nanotechnology to improve the life span of existing consumer goods and create new ones. It seems that they were more concerned with the political macro-aspects than with the consequences of nanotechnology in the everyday life of individuals.

This citizens' conference was only one step in the process to create a public discourse on nanotechnology by the DBT. There have been several follow-ups, but not designed in the same way as this citizens' conference in 2004; and the new initiatives did not involve the same people.

As a result, in 2006, the DBT carried out a classical stakeholders' conference on regulation of health and environmental aspects with nanotechnology processes and products (Teknologirådet, 2006). More than 20 stakeholders took part in the conference and they represented the research community, various public authorities, some businesses, and some NGOs. This stakeholders' conference came up with more specific recommendations than the citizens' conference in 2004. It recommended Danish politicians to extend existing laws and regulations to strengthen Danish activity in international standardisation, to encourage dissemination of risk-related questions, and to establish a research program. Further, it recommended politicians to support public dialogue on nanotechnology, to ensure transparency in the decision-making process, and to apply the precautionary principle generally.

In 2007, the DBT took the initiative to an interactive theatre play on nanotechnology, in cooperation with Centre for Art and Science at the University of Southern Denmark (Teknologirådet, 2007). The play started with a historical overview of the development of science, where nanoscience is one of the latest phases in a dramatic development for mankind. In the sequence the public watched a video discussion among experts on a large screen in the theatre. In the follow-up the audience was invited to take part in the discussion. This three-step design, theatre, video presentation, and public discussion, was repeated two times with slightly different topics. At the very end of the play, the public was invited to answer one question on a postcard: 'Has this play increased your potential to take a personal stand on nanotechnology'? They were also invited to write down some additional comments. These 2006 and 2007 initiatives or events both grew out of the citizens' conference in 2004, as a natural progression of the DBT's engagement with nano.

3.4 Deliberation Criteria

For this last question we will concentrate only on the citizens' conference in 2004. To what extent can this process be regarded as deliberative? By Cohen's (1989) definition, perhaps with some minor reservations, this was a deliberative process:

- It was a *free discourse*; participants regarded themselves as bound solely by the results and preconditions of the deliberation process. However, there might be some more problematic aspects with this conference. First, it was a very short meeting with very limited time and space for discussion among the participants. The main design was a discussion or an interview between DBT and the participants. Secondly, the participants were presented for only two lectures on quite different topics. These two elements probably make it possible for the organisers to influence the outcome.
- It was also a *reasoned* discussion. The participants received a 13-page document before the conference. In addition, they had the opportunity to increase their knowledge on their own before the meeting. At the meeting they were presented with two lectures before they started the discussion in groups.
- The participants were *equal*. They had volunteered to this process, and no one was more equal than the others.
- The conference did not aim at *consensus*; rather the goal was to bring out different voices and opinions.

The strength of this process is that DBT managed to raise serious questions on the relationship between science and society with limited resources. Furthermore, DBT managed to involve ordinary citizens in this process in a way that has been used as a world-wide benchmarking for deliberative processes. In many of the other cases presented in this book we will find reference to the Danish model. The DBT has taken the result from this process into other similar processes, such as the stakeholders' conference. This is also an important positive factor with the way DBT has organised public discussion on nanotechnology.

The resources used in this process were very limited. Three hours may be too little for a deliberative process within this area. It has created an arena for deliberative processes, but it has not taken full advantage of this opportunity. The dialogue between participants seems to be the weak point in this process. The main dialogue was between the organisers and the citizens. We also think that the goals of the meeting were too abstract. The citizens were asked to take specific stands on controversial issues. This may not be a strong argument since the meeting took place in an early stage of the nano-discourse.

What can we learn from this process? First, that it is possible to involve ordinary citizens and consumers in relatively complicated discourses. In that sense, the DBT was very satisfied with the meeting. But more time should be used for the discussion. It might also be fruitful for the participants to meet once more, reopen the discussion, and reflect on the outcomes from the first meeting.

Second, it is also worth noting that even though the DBT has taken several initiatives to raise the question of nanotechnology in the Danish political discourse, this not one of its permanent issues. The board is not running any specific projects post 2007. This means that there might be a need for a more permanent platform for a nanotechnology discourse in Denmark.

3.5 Summarising Appraisal

The citizens' nano conference in Denmark in the summer of 2004 obviously had a number of limitations. They have been accounted for here. The meeting itself was very short, and only one hour of the evening session was used for moderated group discussions. In addition, the introduction to the nano theme was quite modest: a 13-page information package and two lectures. Nevertheless, this small event achieved at least three things:

1. It demonstrated that citizens without prior knowledge of the phenomenon were able to participate in and deliver a meaningful debate on an emerging cutting-edge technology.
2. It, actually, even if unanticipated, produced input to Danish science policy.
3. Finally, and strongly related to the first point, it provided a model for more or less similar processes in Europe and beyond.

There might be some confusion over what the 'Danish model' actually is; is it the use of citizens' deliberations in general, or does it refer to the idea of the consensus conference? We hold the DBT's involvement of ordinary citizens into the democratic processes to be the most correct answer, but references to the consensus conference is not uncommon. We are probably able to live with this lack of conceptual clarity, but we should at least point out that the citizens' conference was *not* a consensus conference.

What is important, however, is that it was the starting point of a wave of deliberative processes over nanotechnologies and over emerging technologies in general.

References

Andersen I-E, Jæger B (1999) Danish participatory models — Scenario workshops and consensus conferences: Towards more democratic decision-making, *Sci Publ Pol,* **26**(5), 331–340.

Cohen J (1989) Deliberation and democratic legitimacy, in *The Good Polity* (Hamlin A, Pettit P, eds), Oxford, Basil Blackwell.

Danish Board of Technology. Available at http://www.tekno.dk/subpage. php3?page=forside.php3&language=uk

Danish Ministry for Science and Technology (2004) Teknologisk fremsyn om dansk nanovidenskab og nanoteknologi. Available at http://www. dtu.dk/upload/centre/nanet/links_ikoner/rapporter%20under%20 links/andre%20rapporter2/teknologiskfremsyn_nanovidenskab_og_ nanoteknologi.pdf

Einsiedel EF, Eastlick DL (2001) Consensus conferences as deliberative democracy: A communications perspective, *Science Communications,* 21, 323–334.

Klüver L (1995) Consensus conferences at the Danish Board of Technology, in *Public Participation in Science: The role of Consensus Conferences in Europe* (Joss S, Durant J, eds), Science Museum, London, pp 41–49. Klüver L (2009) Participatory tools at the Danish board of technology, held at the *Deliberative Processes* Conference in Copenhagen November 10th 2009.

Powell M, Kleinman DL (2008) Building citizen capacities for participation in nanotechnology decision-making: The democratic virtues of the consensus conference model, *Public Understanding of Science,* 17, 329–348.

Teknologirådet (2004) Borgeres holdninger til nanoteknologi (Citizen's attitudes towards nanotechnology), *Citizen Conference* June 7th 2004.

Teknologirådet (2006) Regulering af miljø- og sundhedsaspekter ved nanoteknologiske produkter og processer (Regulation of environmental and health aspects with nanotechnological products and processes), *Stakeholders' Conference* June 2006.

Teknologirådet (2007) *Rapport fra Videnskabsteateret. NANO: Publikums holdninger til nanoteknologi* (Report from the science theatre. NANO: Public attitudes toward nanotechnology).

Chapter 4

The NanoJury in the United Kingdom

Eivind Stø
National Institute for Consumer Research (SIFO), Postboks 4682 Nydalen, 0405 Oslo, Norway
eivind.sto@sifo.no

4.1 Introduction

The NanoJury case in the United Kingdom belongs to what we have come to call the *first generation* of deliberative processes on nanotechnology. The event took place in spring 2005, it lasted for five weeks, and it was reported in September the same year. There were 16 members in the jury, randomly selected, and none of them had qualifications in nanotechnology or in other relevant disciplines.[1]

The concept of a *NanoJury* offers an alternative to many of the other deliberative processes described and discussed in the book. Consensus is one of the criteria used by Cohen (1989) to describe the character of deliberative processes, and consensus conferences are frequently used by the Danish Board of Technology to involve citizens in bottom-up approaches towards technology. However,

---

[1]http://news.bbc.co.uk/2/hi/science/nature/4567241.stm

Consumers and Nanotechnology: Deliberative Processes and Methodologies
Edited by Pål Strandbakken, Gerd Scholl, and Eivind Stø
Copyright © 2021 Jenny Stanford Publishing Pte. Ltd.
ISBN 978-981-4877-61-9 (Hardcover), 978-1-003-15985-8 (eBook)
www.jennystanford.com

NanoJury is not concerned with consensus. The model is inspired by the legal jury system, where the participants reach to verdicts after being presented for relevant information and perspectives from a wide range of different 'witnesses'. Thus, voting is a natural part of the conclusions in such processes.

The NanoJury in Halifax, Yorkshire, was probably the first deliberative process on nanotechnology in the UK. This deliberation was also a part of a broader and deeper U.K. activity on public engagement in complicated and controversial technological questions in the period 2004–2006. There are reasons to believe that the public discourses about bovine spongiform encephalopathy (BSE) and genetically modified (GM) food in the 1990s constituted the political background for this new upstream activity in the United Kingdom, where the government played a decisive part (Gavelin, Wilson, & Doubleday, 2007).

In a document from the U.K. government (HM Government, 2005, p. 2) the goals of a public dialogue on nanotechnology is developed along the following lines:

- Enable *citizens* to understand and reflect on nanotechnology
- Enable the *science community and the public* to explore together aspirations and concerns
- Enable *institutions* to understand, reflect, and respond to public aspirations and concerns
- Establish and *maintain public confidence* in development of technologies
- Contribute to wider *government initiatives* to improve the general trustworthiness
- Support wider governmental initiatives to *support citizens participation*

We have abbreviated the goals slightly and emphasized what we hold to be key phrases in the context of this book.

The U.K. government defined four instruments to meet these goals:

(1) *Nanodialogues* were supported by the Department of Trade and Industry's 'Sciencewise' programme and were developed in cooperation with other projects funded by the Economic and Social Research Council (ESRC) in the United Kingdom.[2]

---

[2] www.sciencewise.org.uk/

Nanodialogues defined four areas for an upstream public engagement, as cooperation with regulatory authorities, research institutions, businesses, and NGOs: risk and regulation in the use of nanoparticles and nanotubes; the role of public engagement in shaping research goals; public engagement in industrial innovation processes; and potential opportunities, barriers, and benefits to the global diffusion of nanotechnology.

(2) The main idea behind *Small Talk* programme was to build an arena for communication between the science community and the public. It was not only a matter of dissemination plans and activities, but also communication strategies and dialogues. To some degree *Small Talk* also represents a necessary link between scientists and political authorities.

(3) The *Nanotechnology Engagement Group* (NEG) (Involve, 2005, 2006, 2007) seems to be the most important of these governmental instruments. NEG consisted of a core research group, a supporting forum of 20 persons, and a wider scientific network. The main goal was to provide a platform for public upstream activity on nanotechnology. Involve was responsible for NEG in cooperation with Cambridge Nanoscience Centre.

The activity of NEG was developed with regard to the governmental goals listed above. In addition, the objectives were to study stakeholders' expectation of public engagement, to map existing public activity, to identify lessons learned, and to develop improved feedback systems and communication programmes.

NEG has formulated a large number of recommendations. These are divided between recommendation to science policy (5) and recommendation for public engagement policy (12)

(4) The fourth instrument included participatory projects not funded by the government. The *NanoJury*, which we are concentrating on in this chapter, was one of these projects.

4.2 Description of the Process

The organisers of this event have taken the concept of a *jury* seriously. The process was organised as a court of law, where the participants constituted the members of a NanoJury. During the process the jury is presented with expert 'witnesses'. The members of the jury may ask questions, and they also have the option to bring in other witnesses to the court. At last the jury reaches a *verdict.*

Institutionally, the NanoJury was an interesting example of a public–private partnership (PPP). It was a cooperative undertaking between Greenpeace, U.K.; the daily newspaper *The Guardian;* and the universities in Cambridge and Newcastle. In addition, two important bodies were established. The first was a scientific advisory board, with multidisciplinary competences to guide the process scientifically. The second was a multi-stakeholder oversight panel to balance the interests of various stakeholders.

The process was organised in four steps. In the first step, the jury was encouraged to discuss local political challenges, not related to nanotechnology at all. They ended up discussing the situation for young people in the community, with the main focus on social exclusion and crime. This slight detour was probably undertaken in order to rehearse the discussion process on a more familiar theme. In addition, participants were able to establish some relations and got a chance to know each other.

In the second step, they turned to nanotechnology. This event took 10 evening sessions of two and half hour each, hence in total 25 hours, where the jury were presented with expert witnesses. These witnesses were picked by the oversight board.

In the third step, the jury formulated possible statements about the future nanotechnology in research and society, and in the last step the participants cast their votes in support for these statements.

This process ended in June 2005, and in the middle of September the final 'verdict' was presented to the public in London. Three jurors attended this concluding event. The main topics were health, jobs, and transparent nanoscience and labels in accessible language.

4.3   Review of the Process

4.3.1   Initiatives and Objectives

The initiative was taken by Greenpeace, U.K. The organisation has earlier been involved in similar activities. In Cambridge its partner was the Nanoscience Centre, and in Newcastle, the Policy, Ethics and Life Science Research Centre (PEALS).

As we have seen, the NanoJury was not directly a part of the government's outline programme for public engagement on nanotechnologies. However, the strategy of the programme was to encourage partnerships in this area and the NanoJury was mentioned under the headline of 'participatory projects not funded by government (HM Government, 2005, p. 11).

Furthermore, the governmental programme established NEG through the Sciencewise grants, and the NanoJury was actually one of the six projects included in the final report from NEG (Gavelin, Wilson, & Doubleday, 2007, pp.13–15). This means that even though the project was not funded directly by the government, the jury was a central part of the governmental strategy. This strategy was not linked directly to specific decision-making processes but was a part of wider strategy to engage the public in the political discourse on emerging nanotechnology.

The NanoJury is described as an upstream public engagement (Sing, 2008). The idea is not to run a process that contributes to the legitimacy of decisions already taken by public authorities but to involve ordinary citizens and consumers in the discourse within a bottom-up approach. To avoid the danger of just co-optation, the process was not only designed for the nanotechnology discourse but also for a more general discussion about emerging technologies. In the opening session, the participants were invited to bring other topics on the table that was essential for their own everyday life in Yorkshire.

The jury came up with several topics, but concentrated the discussions on *young people and exclusion*. This discussion was based upon inputs from youth workers, senior policy officers, and drug-rehabilitation workers. There are reasons to believe that this chosen topic contributed positively to create a framework for the discussion in the next phase of the process; the nanotechnology (Doubleday & Welland, 2007).

4.3.2 Organisation

The NanoJury deliberation was a well-organised process with links to civil society, the scientific community, political authorities, and media. The jury met 18 times in May–July, 2005; meetings took place in the evenings and they lasted for two and half hours each.

It was co-funded by Cambridge University Interdisciplinary Research Collaboration in Nanotechnology; Frontiers Network of Excellence in Nanotechnology; Greenpeace, U.K.; and PEALS at Newcastle University. The direct cost was £45,000 (Gavelin, Wilson, & Doubleday, 2007). This amount covered planning cost and running of the NanoJury process until June 2005. This means that the institutions involved in the process also had to cover substantial cost to finish the NanoJury deliberation.

There were several links between the jury and the civil society. First, Greenpeace was one of the organisers and also took the initiative of the process. Second, a multi-stakeholder oversight panel was established to balance the interests of various stakeholders. This panel played an important part in the selection of topics and 'witnesses' in the jury meetings. Finally, the local society was strongly involved in the first step of the process where participants were encouraged to discuss local political challenges not related to nanotechnology at all. In this phase the participants learned about each other and developed relationships with the organisers of the process.

The link to the scientific community was strong, as two universities participated and organised the process. Furthermore, a scientific advisory board was established, with multidisciplinary competences to guide the process scientifically. Scientists also contributed as expert witnesses in the second part of the process.

The NanoJury was part of a governmental strategy to engage people in complicated scientific discourses. The process was also funded by public money. Even more important was the political recommendation in the last two steps of the process. They were formulated after the witnesses had presented their expert views. The jury then reached its final verdict and presented the verdict at a public meeting in London, where political authorities were also present.

In this last step media played a decisive role. This was not limited to *The Guardian*, which was one of the organisers of the process.

4.3.3 Participation

The NanoJury process was an open discussion, with no limitations and priorities as far as nanotechnologies were concerned. The main idea was the involvement and engagement of citizens in the development of emerging nanotechnologies in the society.

The U.K. citizens' jury was held in the city of Halifax, West Yorkshire. The jury met several times during a five-week period in the spring of 2005. The results were reported in September the same year. This was done on a website in combination with a public event in London (BBC News, 2005). Around 16 to 20 people participated in the jury in the five-week process.

The objectives were formulated in the following way:

- To provide an instrument for people's informed views on nanotechnology
- To facilitate a mutual educative dialogue between people with diverse perspective and interests
- To explore the potentials for deliberative processes to broaden discussion about nanotechnology research policy
- To know about the risks and regulations related to the use of nanoparticles and nanotubes
- To find about the role of public engagement in shaping research goals

In practice the NanoJury concentrated on three topics: health, energy, and ICT, in addition to more general research-related questions. This is reflected in the results and recommendation from the jury.

4.3.4 Results of the Process

The results from the NanoJury process were presented as a 'verdict' at a public meeting in London in September 2007. In phase three of the process the jury formulated various alternative statements about nanotechnology based upon the comments of expert witnesses in phase two. They ended up with 20 statements within the following categories: general statements, health, energy, and ICT (Greenpeace, 2005).

In the last phase the jury threw their votes to these statements. However, they didn't only say *yes* or *no* to the provisional recommendations, they also explained why.

The main topics in the NanoJury discussions were related to the general questions of governance of nanotechnology, openness and transparency, and priorities in the research. These questions cover half of the statements and support was given to five of them and to one with some uncertainty. Let us have a look at some of the examples:

General 3: 'There should be more openness on where public money is spent on nanotechnology research'.

General 7: 'Government should support those nanotechnologies that bring jobs to the UK by investment in education, training, and research'.

General 8: 'Nanotechnologies will only be good if they can enable us to have more quality leisure time including time for families and time for us personally' (Some voters were uncertain).

It is worth noting that the jury did not give strong support to any of the statements dealing with ICT or energy. Two of the ICT statements were 'Normal citizens — people like us — should decide when nanotech starts getting used in ICT' (ICT3) and 'Radiation and other health hazards associated with ICT should be kept low enough so that children can use phones and other ICTs safely' (ICT4). They were both given 'weak support with some voters uncertain'.

Among the three energy statement, strongest support was given to 'Government grants for those pioneering the development, manufacture and use of better solar technologies'. (Energy1). However, the support was given with some uncertainty.

There were two health-related statements, both supported by the jury. They both dealt with testing, safety, and information ('All manufactured nanoparticles should be labelled in plain English...').

It is possible to conclude that the results of this NanoJury reflect the stage in the public discourse on nanotechnology in 2005 and the lack of consumer products on the market. This is probably why the jury did not focus on these matters but gave strong priority to research policy–related questions.

4.4 Deliberation Criteria

In the concluding part of this chapter we will return to the four criteria for deliberative processes developed by Cohen (1989). To which degree has the NanoJury process been (1) free, (2) reasoned, (3) an equal discourse among partners, and (4) aiming at concensus?

As we see it, the way the NanoJury project was developed it addressed three of the four criteria for deliberative processes. The only exception would be the last criterion. It does not seem that there was a strong motivation for a consensus-driving process in the NanoJury. On the contrary, the jury formulated 20 different statements and the members of the jury gave their votes on each statement. This seems to be the logic behind the jury system. In the document this is illustrated by the results of the voting. Some statements are supported, while others only receive weak support, or with some uncertainty. This also indicates that no statement was voted down by the jury. A weak support means that only some of the jury members supported the statement.

It is a *free discourse* in two ways. First, the participants do not represent any specific interests. The composition of the jury broadly reflects the social diversity in the United Kingdom as far as age, gender, education, and social class are concerned, and this diversity is probably crucial for the success of the process. However, we are obviously not talking about any kind of statistical *representativeness*. The organisers also included some religious people in the jury, but they were not representing any specific churches or organisations. They were probably expected to bring in some alternative perspectives on the ethical issues, but this was not a stakeholder conference with representatives for various groups of interests.

Second reason to regard it as a free discourse is that the jury itself chose the topics. It is also worth noting that before the jury started the deliberative nano process, they were invited to discuss another local issue, also chosen by the participants. This strengthened the free dimension of the process. It is possible to argue that the advisory panel and the project group of two universities and one NGO may have had an impact on the discussion climate in the jury. This is also the case for the invited experts in their role as witnesses. They knew far more about the topic than the jury members; but since there was a reasonable balance among the witnesses this does not destroy our conclusion about the free discourse.

The members of the jury were *equal*. None of them had any specific advantages as far as competence in nanotechnology is concerned. If that was the case, they would be excluded. Since the process lasted for several weeks, there are reasons to believe that some of the members of the NanoJury would use the opportunity to improve their knowledge during the process, but this does not challenge the participants' equality. There is, of course, an inequality between the expert witnesses and organisers of the NanoJury on the one hand and the 20 members of the jury on the other, but this difference is not relevant for our main conclusions.

As discussed above this was also a *reasoned* process. A large number of experts from natural sciences, social sciences, and humanities were brought before the jury, representing various perspectives towards the development of nanotechnology. Their main task was to improve the knowledge about nanotechnology among the jury members and help them formulate their dilemmas and statements.

4.5 Summarising Appraisal

One of the important criteria according to Cohen is the question about whether this deliberative exercise was a reasoned process or not. The jury obviously had to reach a verdict on the 20 listed statements, but on what kind of arguments was the conclusion based upon?

As we see it, this was definitely a reasoned process. A large number of experts from natural sciences, social sciences, and humanities were brought before the jury. They gave their presentation of a relatively complicated topic to an audience with no background in natural sciences in general and in nanoscience in specific. This was necessary because the main selection criterion was lack of knowledge of nanotechnology. Such knowledge was the only exclusion criteria from the NanoJury.

In many ways the strength of the jury concept is that the process has to be reasoned. In addition to the presentations, the experts were asked questions by the members of the jury. The jury had also the possibility to propose new witnesses if they wanted to concentrate more on specific topics or they wanted to introduce new perspectives to the discussion. The process also was embedded in the scientific community through the participation of two universities (Cambridge and Newcastle), and a special a multi-stakeholder oversight panel was established to introduce topics and experts for the jury.

This reasoned process is also documented in the recommendation from the NanoJury. In addition to the conclusions, weak or strong support (with or without uncertainty), you will also find some argumentation. The NanoJury supported the statement about bringing jobs to the United Kingdom, as mentioned above. They come up with the following argumentation: 'Without such action, nanotechnologies could lead to greater imports, bringing greater unemployment to the UK' and 'We need to avoid being held hostage to other countries by making sure we invest at the beginning'.

References

BBC News (2005) Citizens' jury to tackle nanotech. Available at http://news. bbc.co.uk/2/hi/science/nature/4567241.stm

Cohen J (1989) Deliberation and democratic legitimacy, in *The Good Polity* (Hamlin A, Pettit P, eds), Oxford, Basil Blackwell.

Doubleday R, Welland M (2007) NanoJury UK. Reflections from the perspective of the IRC in Nanotechnology and FRONTIERS. Available at http://www.frontiers-eu.org/archive/Nanojury%20final%20 reflections%20Mar-07.pdf

Gavelin K, Wilson R, Doubleday R (2007) *Democratic technologies? The final report from the Nanotechnology Engagement Group*, London, Involve.

Greenpeace (2005) Nano Jury UK, Our provisional recommendations. Available at http://www.greenpeace.org.uk

HM Government (2005). *The Government's outline programme for public engagement on nanotechnologies*, August 2005. URN 05/2043.

Involve (2005) *People and Participation: How to put people at the heart of decision-making?* London, Involve.

Involve (2006) *The Nanotechnology Engagement Group Policy Report 1*, London, Involve.Involve (2007) *Putting the Consumer first – Guide*, London, Food Standards Agency.

Sing J (2008) Polluted waters: The UK NanoJury as upstream public engagement. A discussion paper. Available at http://www.nanojury. org.uk/pdfs/polluted_waters.pdf

Chapter 5

Consumer Conference on the Perception of Nanotechnology in the Areas of Food, Cosmetics, and Textiles, Germany

Gerd Scholl

Institute for Ecological Economy Research (IÖW), Potsdamer Str. 105, 10785 Berlin, Germany

gerd.scholl@ioew.de

5.1 Introduction

The German Consumer Conference on the perception of nanotechnology in the areas of foodstuffs, cosmetics, and textiles was a process of public deliberation conducted in 2006. It was initiated by a public authority (German Federal Institute for Risk Assessment) as part of its risk communication activities. Since it was the first of its kind run by this authority, the Consumer Conference constituted a social experiment aiming to explore the opportunities of public deliberations. As a result the process provided a consumer vote containing recommendations on how to deal with nanotechnologies in the selected consumption domains.

Consumers and Nanotechnology: Deliberative Processes and Methodologies
Edited by Pål Strandbakken, Gerd Scholl, and Eivind Stø
Copyright © 2021 Jenny Stanford Publishing Pte. Ltd.
ISBN 978-981-4877-61-9 (Hardcover), 978-1-003-15985-8 (eBook)
www.jennystanford.com

5.2   Description of the Process

The Consumer Conference was launched as a pilot project by the German Federal Institute for Risk Assessment (BfR). The process started in March 2006 and was finalised in January 2007. Methodologically, the Consumer Conference refers to the Danish model of the consensus conference.[1] The main objective of this tool of civil participation is to assess new scientific or technological developments from the point of view of informed citizens. It is characterised by a structured and open dialogue between laymen and experts. Before the Consumer Conference on nanotechnologies, three similar conferences — also referring to the Danish model of public deliberation — had been conducted in Germany (on genetic diagnostics, stem cell research, and brain research); none of them, however, was run by the BfR.

Hence, with the Consumer Conference on nanotechnologies it was for the first time in Germany that a public institution employed such an instrument of public deliberation. According to the president of the BfR, the conference was designed to reveal the different perspectives, judgements, and expectations of the consumer group (Hensel, 2008). It was launched as part of a comprehensive research strategy identifying potential risks of nanotechnologies. The strategy comprised, amongst others, a Delphi survey among nanotechnology experts and a representative consumer survey (Zimmer *et al.*, 2008a) as well as an analysis of the media coverage of nanotechnologies.[2] All these activities were to generate orientation and to enhance social capacity to act in a rapidly emerging field of technology (Hensel, 2008).

In particular, the *objectives* of the Consumer Conference were as follows:

- To overcome information deficits and promote a differentiated opinion-forming process on nanotechnology amongst consumers

---

[1]See   www.tekno.dk/subpage.php3?article=468&toppic=kategori12&language=uk for a definition from the Danish Board of Technology, and Durant and Joss (1995) for a broader discussion of tools for public participation in science and technology including the consensus conference.

[2]This study was conducted to explore how nanotechnology is discussed in German mass media, particularly, newspapers and news magazines (cf. Zimmer *et al.*, 2010).

- To prepare an informed vote by consumers on applications of nanotechnology in the areas of foodstuffs, cosmetics, and textiles
- To advise decision-makers by delivering the consumer vote to representatives of policy, science, business, and civil society (Zimmer *et al.*, 2008b).

The focus on foodstuff, cosmetics, and textiles was chosen by the BfR, since it observed increasing application of nanoparticles and nanomaterials in these consumption areas (Zimmer *et al.*, 2007).

For organisation and management of the process, the BfR commissioned two institutes: the Independent Institute for Environmental Concerns (UfU), which provided expertise in citizen participation procedures, and the Institute for Ecological Economy Research (IÖW), which contributed by its expertise in nanotechnologies. The process itself comprised two preparatory weekends with the lay panel and a third weekend during which an expert hearing and the formulation of the consumer vote were conducted.

The *timeline* of the Consumer Conference was as follows:

- April 2006: Setting up of the project, definition of the approach
- May 2006: Establishment of a scientific advisory board; mailing of invitations to randomly selected citizens
- June 2006: Appointment of a moderator; drawing of the participants from the sample of citizens who had responded to the mailing
- July 2006: Identification of possible experts for the public hearing
- August 2006: Conceptual and organisational prearrangement of the two preparatory weekends with the consumer group
- September 2006: First preparatory weekend with the consumer group plus follow-up
- October 2006: Second preparatory weekend with the consumer group plus follow-up
- November 2006: Public expert hearing, preparation of the consumer vote
- December 2006: Dissemination activities
- January 2007: Final reporting

The scientific *advisory board* consisted of four academics experienced in nanotechnologies and nanoscience and also in risk management and risk communication. The board advised the organisers of the conference in content-related and procedural matters.

An essential component of a consumer conference is the selection of the lay panel. A mailing was sent to the 5750 randomly chosen citizens of a confined area[3] asking for their interest to participate in a consumer conference on nanotechnologies. Forty-one letters were returned corresponding to a return quota of 0.71%. This low quota may be due to the fact that nanoscience and nanotechnologies were not on people's minds at that time and/or were not perceived a controversial issue (Zimmer *et al.*, 2008b, p. 10). The 16 members of the lay panel were randomly drawn from this group of 41 citizens, which was sorted along sex and age beforehand. A precondition for participation was that no professional interest in nanotechnologies existed. The panel covered seven women and nine men from 20 to 72 years of age, all with different backgrounds, a student and a young mother, an unemployed and a self-employed, a financial accountant and a managing director, and a retired teacher and a retired farmer. The participants were reimbursed all costs for travelling and accommodation. They did not receive any additional remuneration.

The process of the Consumer Conference was split into the following steps: The panel received introductory information material on nanotechnologies and on the parties involved in the process prior to the first meeting. The content of the info package was cleared between the process organisers, the scientific advisory board and the BfR. During the first preparatory weekend the group was given an expert input (presentation plus questions and answers) and elaborated a preliminary list with questions on nanotechnologies in working groups and plenary sessions.

- At the second weekend, two more inputs were provided, one by a science journalist and one by a representative from the cosmetics industry, and discussed in the plenary. Afterwards, the lay panel finalised the list of questions and selected the experts for the hearing from various stakeholder groups (science, associations, public agencies, industry).
- The closing event was held in Berlin from 18 to 20 November 2006. During this third weekend, the panel consulted the

---

[3]The geographical limitation to the Berlin/Brandenburg area was to enable a cost- and time-efficient organisation of the conference while at the same time ensuring a broad enough socio-demographic scope of the citizen sample.

invited experts along the list of questions (half a day per product domain) and prepared its consumer vote. The preparation was split into three working groups and started Sunday afternoon. The final draft of the vote was eventually adjusted in a moderated plenary session of the lay panel ending around midnight the same day. On Monday morning the vote which was supported by all members of the lay panel was presented to the public and handed out to representatives of the BfR, the Federal Ministry of Food, Agriculture, and Consumer Protection (BMELV), the committee on 'food, agriculture, and consumer protection', the German Parliament, and an environmental NGO ('BUND').[4]

- The vote formulated by the lay panel calls, amongst others, for comprehensible labelling, clear definitions, terms and standards, as well as far more research into the potential risks before nanotechnology is used to a larger extent in consumer products. The vote names foodstuffs as the most sensitive area for the use of nanomaterials. The group felt that the promised advantages to be derived from using nanotechnology like changes to the flow properties of ketchup or the trickling properties of products were non-essential given the potential risks. Regarding the use of nanotechnology in cosmetics and textiles the consumers felt that the already foreseeable benefits clearly outweighed potential risks.[5]

The budget provided by the BfR to conduct the Consumer Conference this way accounted for approximately 100,000 euros, two-thirds of which covered personal costs at the two institutes organising the ten-month process. The rest were material costs, most of it for catering as well as travel and accommodation for the lay panel and experts and rent for the conference rooms.

The Consumer Conference was evaluated in retrospect. From July to October 2007 additional research was commissioned by the BfR aiming to assess the process against a number of criteria (Zimmer *et al.* 2008b, p. 34f):

- Effectiveness (Were the goals set actually attained?)

---

[4]Remarkably, the organisers were not successful in inviting representatives from business and science to this public event.
[5]See http://www.bfr.bund.de/cm/245/bfr_consumer_conference_on_ nanotechnology_in_foods_cosmetics_and_textiles.pdf for an English version of the consumer vote.

- Efficiency (Were the goals attained with justifiable efforts?)
- Fairness (Were all interests treated in a balanced way?)
- Competence (Were all important subjects touched and was scientific evidence considered appropriately?)
- Transparency (Were the procedures clear to everybody involved? Were the external observers able to get a picture of the process?)

The evaluation was based on interviews with members of the lay panel, the advisory board, the group of experts, and the organising institutes. In addition, a follow-up meeting with the consumer group was organised in December 2007, during which the organisers and initiators of the Consumer Conference provided feedback on the impact of the Consumer Conference and its follow-up.

5.3 Review of the Process

5.3.1 Initiatives and Objectives

The Consumer Conference was not part of a distinct decision-making process. The initiator, a sub-ordinate agency to the Federal Ministry of Consumer Protection, dealing with risk assessment and risk communication, had launched the process in order to reveal consumers' risk perceptions at an early stage of technology development. As mentioned above, the agency intended to strengthen the policymakers' ability to act and by doing so followed a kind of precautionary principle.

At the time the conference was conducted, technology governance was at an early, albeit quite visible stage. Several studies had been commissioned addressing the chances and risks of nanotechnologies for humans and the environment (Steinfeldt *et al.*, 2003, 2004; UBA, 2006) and the ensuing regulatory demands (Führ *et al.*, 2007). A research strategy on the risks of nanotechnologies on health and the environment was developed in parallel to the Consumer Conference.[6] And about one year before the conference started, the case of health disorders caused by allegedly nano-based sealing

---

[6]It was published in December 2007 and represents a joint effort of the BfR, the Federal Institute for Occupational Safety and Health (BAUA), and the Federal Environmental Agency (UBA) (BAUA *et al.*, 2007).

sprays had alarmed a broader public.[7] Thus, the deliberative process was embedded in a societal environment that already grappled with the new technology in numerous ways.

Since the officially stated objectives of the Consumer Conference were rather general, one may conclude that the process was at best to some extent intended to give advice on how to proceed further with nanotechnologies in consumer applications. However, the fact that the BfR commissioned an ex-post evaluation of the process and organised a follow-up meeting with the consumer group indicate that it took the pilot very seriously. This impression is confirmed by a press release of the BfR in which it concludes that 'felt risks are part of society and coin people's everyday lives. Hence, in order to avoid crises governmental action is required also in case of felt risks. (...) Communication instruments such as the consumer conference contribute to tracing back felt risks to their rational, scientifically provable core.' (Zimmer, 2007).

Hence, from a general perspective one can argue that the Consumer Conference — as a unique experiment in participation-based risk communication — has reached its goals. Content wise, however, the ex-post evaluation paints a somewhat more-differentiated picture: Although the evaluators conclude that the Consumer Conference provided good results within a limited monetary and timely budget, they doubt whether the consumer vote, which was the main outcome of the process, vicariously reflects consumers' perceptions of nanotechnologies in the selected domains (*ibid.* 35). Instead, they argue for multiple citizens' panels conducted in parallel and providing a broader range of views and deliberations. From this perspective, one can come to the conclusion that the deliberative process was under-scaled in terms of providing a more representative citizen–consumer voice.

5.3.2 Organisation

According to the evaluation, the process was equipped with a relatively small amount of monetary resources but can be regarded

---

[7]According to the findings of the Federal Institute for Risk Assessment (BfR), nanoparticles were not the cause of the health disorders which occurred after using the sealing sprays under consideration. Based on the information from manufacturers and chemical studies commissioned by BfR, the products did not contain any nanosized particles. The term 'nano' in the product names ('Nano Magic') was intended far more to draw attention to the wafer-thin film that forms on the surface of glass or ceramic after the spray-application of the products.

as a 'particularly cost efficient way of civil participation' (Zimmer *et al.*, 2008b, p. 40). This was at least partly due to the high level of commitment of the laymen. Altogether, the citizen-consumers spent, voluntarily, three weekends directly in process events and also some additional time of private deliberation in between the face-to-face meetings. This substantial amount of time allowed for a more thorough appropriation of scientific knowledge and for in-depth reflections of the chances and risks of nanotechnologies in the three domains — obviously an advantage of such small-scale deliberation.

5.3.3 Participation

Numerous stakeholders were involved in the process in different ways:

- BfR was involved as an initiator and funder of the process
- Sixteen citizen-consumers formed the lay panel, matching different socio-demographic requirements (see above).
- Four members of the scientific advisory board with different disciplinary backgrounds provided comments to the process design and, moreover, enhanced the legitimacy of the deliberative process, for instance, vis-à-vis the scientific community.
- Three experts provided inputs during the two preparatory weekends; one from the side of the organisers, one science journalist, and one representative from an industrial association (IKW — the German Cosmetic, Toiletry, Perfumery, and Detergent Association).
- Thirteen experts provided answers to the questions from the lay panel during the public expert hearing. They are spread among science and research (8), federal offices (2), civil society organisations (2), and industry (1).
- Four people from the two institutes organised the process, being in charge of all administrative tasks and moderation of the preparatory weekends plus one extra moderator for the final weekend (facilitating the expert hearing and the preparation of the consumer vote).

As mentioned above, the lay panel was drawn from a random sample. This random sample, however, was confined to a specific region. From the point of view of the advisory board, however, this did not generate a significant bias in the results, but rather ensured a cost-efficient accomplishment of the process (Zimmer *et al.*, 2008b, p. 29).

The scope of the Consumer Conference was defined beforehand. The limitation to cosmetics, textiles, and food was perceived as an advantage, since it enabled a focussed and efficient debate of the issues (Zimmer *et al.*, 2008b, p. 30). The consumer group particularly appreciated the factual justification of this selection that was given by a representative of the BfR at the beginning of the first preparatory weekend. Although side issues, such as military applications of nanotechnologies, popped up at times in the process, the lay panel did not encounter severe problems in re-focussing the discussion again.

With respect to the expert panel, academic expertise was over-represented in the process, while the perspective of industry and CSOs was far less pronounced. This was, on the one hand, due to the fact that representatives from business — particularly in the food domain — were less willing to make themselves available for the expert hearing and, on the other hand, a result of how the lay panel selected people from the list of possible experts (which included three company representatives).[8] The same holds for the stakeholders from civil society, such as consumer or environmental organisations. However, the list of possible experts was much shorter in this area. Besides that, recruiting experts in general proved to be a time-consuming and intricate task, since plenty of scientific domains had to be covered, many people had to get in touch with, and lots of information had to be compiled and processed in a way easy to digest for the lay panel.

The evaluation of the Consumer Conference concludes as to the fairness of the process and particularly the consumers' involvement that each consumer was given the right to articulate him- or herself at any time in the process and that single opinions were appropriately taken into account. It criticizes, however, that the group of consumers was fairly small, and not all citizens possibly concerned could take part in opinion forming (Zimmer *et al.*, 2008b, p. 35).

---

[8]During the expert hearing, 120 people were invited to serve as experts. Fifty of them returned a positive response. Just before the second preparatory weekend, they were asked to send a brief profile as the basis the lay panel should choose their experts from. Thirty experts did so.

5.3.4 Results of the Process

The major outcome of this deliberative process was the consumer vote which was adopted unanimously by the lay panel. The vote contains recommendations for science, business, and policymaking on how to deal with nanotechnologies in the selected consumption domains. In this respect, one can characterise the Consumer Conference as a process aiming at the representation of the common voice, rather than just being a process involving stakeholders, who represent the interests of their various constituencies.

The BfR took a number of initiatives to disseminate the consumer vote among decision-makers (Zimmer, 2007). They presented the vote on scientific conferences, at the German 'Nano-Commission' (a multi-stakeholder board), at the consumer committee of the German Bundestag, at Federal and regional authorities, at industrial associations, and at the European Food Safety Authority (EFSA). Information as to if or to what extent these promotional activities actually had an impact on decision-making is not available though. At least, the follow-up meeting with the lay panel of the Consumer Conference that was held one year after the publication of the consumer vote provided some room for critical reflections of the overall outcomes.

Besides the consumer vote as such, the deliberate process resulted in a kind of public opinion-building process en miniature. Members of the lay panel reported that they brought the topics of the Consumer Conference into their daily lives, for instance, in talks with family members and friends in between preparatory weekends. Thereby, they intensified their own deliberations and contributed to an informal social dissemination.

5.4 Deliberation Criteria

All in all, the Consumer Conference can be regarded a free discourse. All thematic restrictions were properly explained and justified. It can be regarded a reasoned process also. There were diverse means of interaction and face-to-face communication applied. The main touch points where scientific argument was fed into the process were the two preparatory weekends and especially the expert hearing. The ex-post evaluation tells that the experts regard the consumer vote as comprehensive and sufficiently reflecting the scientific state-of-the-art. The variety of opinions included in the process, however, suffers from a lack of industry voices — for the reasons mentioned. If industry were involved more intensely in the process of the Consumer Conference, the group of laymen might have come up with a somewhat different — maybe more, maybe less critical — consumer vote. Besides that, the members of the lay panel were attested appropriate communication skills enabling their equitable participation in the debates (Zimmer *et al.*, 2008b, p. 36).

Mutual learning, particularly among the lay panel, was stimulated by altering compositions of working groups and also by the fact that in between the three weekends there was plenty of time to digest the discussions and the information obtained. Moreover, one of the industry representatives who attended the second preparatory weekend told that her participation in the discussion with the lay panel provided several hints on how to better communicate with the general public on issues potentially perceived as bearing some risk.

Hence, the evaluation arrives at the conclusion that the Consumer Conference was, by and large, a transparent process of deliberation. Consumers and experts had a clear understanding of their roles and the entire process was made well traceable for outsiders through an extensive press work (Zimmer *et al.*, 2008b, p. 36). This impression is underpinned by the fact that all procedural steps are well documented in the final report of the Consumer Conference. This report contains, for example, the minutes of every preparatory weekend that had been circulated to the lay panel, the advisory board, and the BfR.

5.5 Summarising Appraisal

The German Consumer Conference on nanotechnologies can be regarded as a typical process of public deliberation and civil participation. Its strengths are its well-defined scope, its high level of transparency, its proper management of multi-stakeholder participation, and its good cost–benefit ratio. The process scores less good, however, on the fairness scale: It is questionable to what extent the consumer vote can be seen as a voice of the German citizen-consumer in general.

Besides this limitation and going through the analysis by Zimmer (2007), one can learn from this deliberative process that consumer conferences can

- deliver competent and comprehensible results in a cost-efficient way
- portray a multilayer picture of citizens' perceptions of technologies
- provide hints on the possible social risks of nanosciences and nanotechnologies
- reflect how consumers weigh different (perceived) risks
- contribute to building trust into public risk assessment and management

The overall effectiveness of such a process is highly sensitive to proper timing of its different stages, leaving enough room for sound preparation and private deliberation in between meetings and, moreover, to a thorough integration of expert knowledge. The prospects of consumer conferences will depend upon exploring further potentials for cost reductions that may, if exploited, allow for a broader application of this tool of public engagement.

References

BAUA, BfR, UBA (2007) Nanotechnologie: Gesundheits- und Umweltrisiken von Nanomaterialien – Forschungsstrategie *(Nanotechnology: Health and environmental risks - research strategy)*, n.p.

Durant J, Joss S. (1995) *Public participation in science. The role of consensus conferences in Europe.* London: Science Museum.

Führ M, *et al.* (2007) Rechtsgutachten Nano-Technologien *(legal appraisal of nano technologies)*, Dessau. Available at http://www.umweltdaten.de/publikationen/fpdf-l/3198.pdf

Hensel A (2008) Vorwort *(Preface)*, in *BfR-Verbraucherkonferenz Nanotechnologie. Modellprojekt zur Erfassung der Risikowahrnehmung bei Verbrauchern (BfR Consumer Conference nanotechnologies. Pilot project on the risk perception of consumers)* (Zimmer *et al.*, eds), Berlin.

Steinfeldt M., Petschow U., Haum R., von Gleich A. (2004) Nanotechnology and Sustainability, Discussion paper IÖW 65/04, Berlin.

Steinfeldt M, Petschow U, Hirschl B (2003) Anwendungspotenziale nanotechnologiebasierter Materialien. Analyse ökologischer, sozialer und rechtlicher Aspekte *(Application potentials for nano materials. Analysis of ecological, social and legal aspect)*, Publication series of IÖW 169/03, Berlin.UBA (2006) Nanotechnik — Chancen und Risiken für Mensch und Umwelt, Hintergrundpapier *(Nanotechnology —*

Chances and risks for man and the environment. Background paper), August 2006, Dessau. Available at http://umweltbundesamt.de/uba-info-presse/hintergrund/nanotechnik.pdf

Zimmer, R. (2007) Die Bewertung der Verbraucherkonferenz und ihre politische Bedeutung *(The assessment of the Consumer Conference and its political implications)*, presentation at the follow-up meeting of the lay panel, Berlin.

Zimmer R, Domasch S, Scholl G, Zschiesche M, Petschow U, Hertel R, Böl G-B (2007) Nanotechnologien im öffentlichen Diskurs. Verbraucherkonferenz zur Nanotechnologie *(Nanotechnologies in public discourse. Consumer Conference on nanotechnologies)*, in: *Technikfolgenabschätzung. Theorie und Praxis*, No. 3/2007, 16. Jg., pp. 98–101.

Zimmer R, Hertel R, Böl G-B (Hrsg.) (2008a) Wahrnehmung der Nanotechnologie in der Bevölkerung, Repräsentativerhebung und morphologisch-psychologische Grundlagenstudie *(Perception of nanotechnologies in the public, representative survey and morphological-psychological baseline study)*, Berlin. Available at http://www.bfr.bund.de/cm/238/wahrnehmung_der_nanotechnologie_in_der_bevoelkerung.pdf

Zimmer R, Hertel R, Böl G-B (Hrsg.) (2008b) BfR-Verbraucherkonferenz Nanotechnologie. Modellprojekt zur Erfassung der Risikowahrnehmung bei Verbrauchern *(BfR Consumer Conference nanotechnologies. Pilot project on the risk perception of consumers)*, Berlin. Available at http://www.bfr.bund.de/cm/238/bfr_verbraucherkonferenz_nanotechnologie.pdf

Zimmer R, Hertel R, Böl G-B. (Hrsg.) (2010) Risk Perception of Nanotechnology — Analysis of Media Coverage, Berlin. Available at http://www.bfr.bund.de/cm/238/risk_perception_of_nanotechnology_analysis_of_media_coverage.pdf

Chapter 6

French Conferences Cycle on Nanotechnology: Nanomonde

Giampiero Pitisci
Independent Designer
giampiero.pitisci@solutioning-design.net

6.1 Introduction

The French 'Nanomonde' was a process of public deliberation conducted in 2006. It was initiated by a French NGO as part of its 'vigilance on democracy' activities and provided 17 recommendations on how to deal with nanotechnologies for selected domains.[1]

[1]Nanomonde official website: http://www.vivagora.org/spip.php?rubrique31. Nanomonde Advisory board: Bernadette Bensaude-Vincent, Philosopher and Historian of Sciences, Université Paris X, Nanterre/Philippe Bourlitio, civil organisation Sciences et démocratie/ Sylvie Catellin, Laboratoire Communication & Politique, CNRS/Alain Lombard, toxicologist, ex-Arkema/Francis Chateauraynaud, sociologist, EHESS/Claudia Neubauer, coordinator of Fondation Sciences citoyennes/Ariel Levenson, responsible of the Centre de compétences en nanosciences d'Île-de-France, C'Nano/Roger Moret, C'Nano/Véronique Thierry-Mieg, ECRIN/Eric Charikane, ECRIN/Claude Henry, sociologist, Président de VECAM/Louis Laurent, responsible of the sector Materials and Informations, Agence nationale de la recherche/Gérard Toulouse, physician and member of the Comiéte permanent sur Sciences & éthique d'ALLEA (Alliance européenne des académies)/Françoise Roure, Treasury, Conseil général des technologies de l'information/Chloé Ozanne, student in communication, Grenoble/Mohamed Belhorma, CCSTI Grenoble/Arnaud Apoteker, Greenpeace/Roland Schaer, responsible for 'Collège relation Sciences et société', Cité des sciences et de l'industrie.

Consumers and Nanotechnology: Deliberative Processes and Methodologies
Edited by Pål Strandbakken, Gerd Scholl, and Eivind Stø
Copyright © 2021 Jenny Stanford Publishing Pte. Ltd.
ISBN 978-981-4877-61-9 (Hardcover), 978-1-003-15985-8 (eBook)
www.jennystanford.com

6.2   Description of the Process

Nanomonde was organised by Vivagora, a Paris-based French NGO founded in 2003. Vivagora frequently organises large public debates on social issues related to scientific and technological developments. It aims at encouraging interactions between academic, public, industrial, and civil society actors. Its already organised conferences are 'The Appropriation of the Living' (2004), 'Health and Environment' (2005), and 'Brain and Mental Health' (2007).

In addition Vivagora, each year, in collaboration with the Conseil régional Île-de-France, organises a meeting on 'Innovations in Democracy' — the latest was held in March 2008.[2]

The Vivagora website was conceived as an online 'vigilance' tool on participatory democracy techniques and social aspects of emerging technologies. It publishes a quarterly letter (*VivaAgoVeille*) to inform on these and related issues. Aside from the organisation of public debates, Vivagora provides a broad range of information on sustainability, neurosciences, and nanotechnology.

The Nanomonde cycle was the first of a three nano-related series of conferences held in France between 2006 and 2007 (Nanomonde Paris 2006, Nanoviv Grenoble 2006, Nanoforum Paris 2007). A cycle of six conferences was launched in 2006. The process started in January 2006 and ended in June 2006. Seventeen recommendations were presented to and discussed with the public on the sixth and final conference. Methodologically, the Nanomonde conference explicitly refers to 'national initiatives in public debate on nanotechnology like the Danish Board of Technologies (DBT) (2004, see Chapter 3), the citizen conference organised by the University of Wisconsin (Madison, April 2005, see Chapter 9) and the English NanoJury (summer 2005, see Chapter 4).[3]

Nanomonde aimed at assessing new scientific or technological developments from a social point of view, co-constructed in an open dialogue between public and experts. Its main objective could be formulated as advising decision-making by delivering citizens vote to academic, industrial, political, and civil society actors.[4]

---

[2]http://www.vivagora.org/spip.php?rubrique66
[3]http://www.vivagora.org/spip.php?rubrique31
[4]See final report with recommendations: http://www.vivagora.org/IMG/pdf/Pr%
E9conisNANOMONDE%20oct06DEF.pdf (p. 1)

The Nanomonde cycle adopted a generalist position regarding nanotechnologies and society. The themes were from the first conference to the sixth:

1. What's new with nanotechnologies?
2. Between Science and Fiction: Actors, discourses, and issues
3. Nanotechnologies and energy: Towards sustainable developments
4. Communication and transport: What kind of usage for daily nanoproducts?
5. Nanotechnologies and health: Hopes, norms, responsibilities, and risks
6. Nanoperspectives: Geopolitics, military, and democratic issues.

The third and fourth conferences were more than the rest aimed at consumers' issues like energy use and mobility. Whether health is a consumer issue or not is open to debate, and Europeans and Americans might have different views on this.

Nanomonde took place in the 'Cité Universitaire de Paris', a private foundation of public utility. Collaborative partners mainly came from the press/media sector and from civil society organisations: France Inter, Le Monde, Alternatives Économiques, Sciences & Avenir, Valeurs Vertes, Vivantinfo website, Université Paris-Sud, Commission éthique de l'Académie des technologies, Centre de compétences Nanosciences en Île-de-France (C'Nano), Association ÉCRIN (CNRS-Industries), Association ORÉE industrials for environment), Association Sciences et Démocratie.

As mentioned above, the Nanomonde cycle started in January 2006, with one conference per month for six months. The use of time for setting up the project, for organisational arrangements, and for identifying final experts has unfortunately not been communicated.

The scientific advisory board consisted of seventeen members with different disciplinary backgrounds (see the list at the end of the chapter). The board advised the organisers of the conference in content-related and procedural matters.

Each conference consisted of 100–120 citizen participants. For them there was no pre-selection, and participation was free and open. In addition a number of organisers, experts, and stakeholders were attending.

In short, the core Nanomonde process can be described as follows: six public conferences, one per month for six months. The public was informed of the conference through the Vivagora website and the national press. A few weeks before each conference, an introductory paper (Fiche-Repère) of two or three pages was available online. It aimed at providing technical information about the conference theme. Each conference lasted for about three hours; from 7 pm to 10 pm. Minutes of each conference became available online within eight days.

It seems that no external evaluation of the Nanomonde deliberative process has been made. An internal evaluation was made during the process by using a classical questionnaire. Unfortunately, no significant conclusions can be drawn regarding the public feedback because of the low number of answers — less than 10% of the participants answered the questionnaire.

Some considerations after the two first conferences can be found on the Vivagora website.[5] The author, one of the advisory board members, stressed the difficulties to reach visibility in clarifying definitions and ethical issues and in trying to think 'disagreement within unity'.

6.3 Review of the Process

6.3.1 Initiatives and Objectives

Before the Nanomonde cycle, it seems that no major attempts of utilizing deliberative processes had been made in France in the nanotechnology/nanoscience field. It is important to note that the well-known 'Conférence citoyenne sur les nanotechnologies' (see Chapter 7) took place in Paris a couple of months later, from October 2006 (preparatory sessions) to January 2007 (the conference itself); with some of the same members in the scientific advisory board[6] (see Chapter 7). The initiative came from Vivagora, and the explicit objective was to introduce an element of participatory

---

[5]http://www.vivagora.org/spip.php?article76
[6]For more information on the Conférence citoyenne sur les nanotechnologies, see http://www.enroweb.com/blogsciences/index.php?2006/10/18/63-conference-de-citoyens-sur-les-nanotechnologies

democracy into the social and ethical aspects of the emerging field of nanoscience and nanotechnology.

6.3.2 Organisation

The Nanomonde process was equipped with a rather small amount of money; 9000 euros from the Conseil Régional Île-de-France — and it was made possible; thanks to the collaboration from Cité Universitaire de Paris'. The Cite provided conference rooms and technical support. The budget mainly covered material costs, mainly of catering plus travel costs and accommodation for experts.

As mentioned above the content was communicated beforehand via introductory papers to public participants. These papers came as rather short presentations of the conference themes, with bibliographical and Internet sources. Each introductory paper was of about two or three pages. Because of the lack of external evaluation of the process, we do not know if participants found these papers to be useful tools for getting involved in a nanotechnology debate. Neither do we know if they preferred the generalist tendency of the first conference cycle to the second one, Nanoviv in Grenoble, which focused more specifically on the relation between nano- and biotechnologies.

We don't have any significant feedback from public participants — less than 10% of answered questionnaires. As a contrast, in the latest Vivagora's Nanoforum (2008), 90% — of the 80% who returned the questionnaires — responded positively. Questions were mainly about how they assessed the quality of information and whether they felt to have been part of a constructive and meaningful participatory process.

6.3.3 Participation

There were as many as between 100 and 120 citizens attending each conference. They participated in their role as interested, self-selected citizens. We do not have access to information about the socio-demographic profiles of the participants but one can suppose they all belong to the same geographical area (Paris). With such a big attendance the conferences obviously were very different from the ones performed more in the 'focus group format', and perhaps appeared more to be a kind of political mass meeting.

In addition to the lay participants, a set of different stakeholders were involved in the process:

- Vivagora, as initiator and organiser of the process; Cité Universitaire de Paris as the host of the conferences
- Seventeen members of the scientific advisory board with different disciplinary backgrounds
- The figure of a 'witness' (Grand Témoin) — in this case Dominique Pestre, Physician an Historian of Sciences (EHESS) — in charge of 'pushing on the controversies, linking interventions and clarifying their meanings, summarizing the debate, and pointing out elements for a common deliberation'[7]
- An animator/moderator from Vivagora
- A resource base of 34 experts for the entire process (about 6 for each conference) provided answers to the questions from the public. Of these, 9 came from the natural sciences and/or technology background, 10 from the social sciences and economy background, 6 from industry, 4 from public authority, 3 civil society organisations, and 1 was a writer and 1 filmmaker.

It is important to note that each conference's expert panel aimed at being balanced, reflecting sometimes contradictory interests from industry, civil society organisations, or science researchers. The experts did not prepare formal presentations; they just answered public questions. In effect, the conference started as a debate conducted by the moderator and the 'witnesses'. Each conference was based on a three-step structure:

1. Identification and formulation of the issues
2. Identification of conflicts
3. Deliberation, that is, formulation of suggestions and possible solutions

No precise information has been communicated regarding the time allowed for each step. Public participants could ask questions or formulate comments at every moment.

---

[7]http://www.vivagora.org/spip.php?rubrique44

6.3.4 Results of the Process

The main objective was to help in decision-making by delivering citizens votes to 'academic, industrial, political, and civil society actors'. Vivagora formulated 17 recommendations after the cycle of conferences, and particularly insisted on the co-elaborative nature of these recommendations. These were presented and discussed with the public during the sixth conference (June 2006). Among other things, Vivagora pointed out that the conferences really did not show a public fear of nanotechnologies, but rather a lack of trust in 'those people who make science'.[8] Academic actors where invited to promote a more extensive public information culture about science; industry actors were asked for more transparency; and political actors were asked to promote a European agency of vigilance on new technologies.

Unfortunately, Vivagora did not provide a precise list of actors to whom these recommendations were addressed. The organisers considered, however, that these recommendations actually had *no impact on decision-making* whatsoever. According to them, this was mainly due to the fact that Vivagora was not commissioned by a public authority and did not belong to any academic institution. Regarding this, the next Nanoviv (Grenoble, 2006) was organised differently: it was commissioned by the Region Rhône-Alpes, so the political engagement in the cycle was publicly stated. However, even for this case, still according to organisers, the real political impact of recommendations formulated during the process was very poor.

The fact that Vivagora organised two other conferences (Nanoviv and Nanoforum) indicates that they regarded this form of deliberative process as a valuable tool for getting public participation in new democratic techniques. It is important to note that the third cycle of conferences — Nanoforum (Paris) — *did not result in any recommendations* despite its large audience and apparent success. In effect, Vivagora is currently preparing a new series of conferences based more on a 'co-construction of questioning' than on obtaining recommendations. This new version of public deliberation process will be first dedicated to 'synthetic biology'.[9]

---

[8]See final report with recommendations: http://www.vivagora.org/IMG/pdf/Pr%E9conisNANOMONDE%20oct06DEF.pdf
[9]First cycle was performed in 2009 in Paris.

6.4 Deliberation Criteria

We will consider Cohen's criteria: was it a free process? Was it a reasoned process? Were participants equal? Did it aim at rational consensus?

The process was free, meaning that the participating citizens themselves defined, formulated, and identified the issues; considered possible conflicts; and formulated suggestions and solutions. We do not know the precise role of the moderator and/or of the witnesses, but we might suspect that the voice of the experts will be influential is such large conferences conducted in limited time.

The Nanomonde conferences were conducted as reasoned processes, as open dialogues between public and experts. As depicted above, the deliberative process employed 'classical' means of interaction and face-to-face communication. Neither specific working groups nor scenario techniques were used.

For the participants, the lay citizens, it seems as if equality was sought and achieved. A bit beyond Cohen's criteria, one might, however, question the equality between participants and experts, but that is a more general question, perhaps relevant for all the citizen-oriented deliberations reviewed here.

The conference cycle was apparently not concerned with consensus. Instead it was voted over the different recommendations.

6.5 Summarising Appraisal

Nanomonde aimed at assessing new scientific or technological developments from a social point of view co-constructed in an open dialogue between public and experts. Its main objective was advising decision-making by delivering citizens' vote to academic, industrial, political, and civil society actors.

Despite the high number of participants and the rather short time allowed for debate (approximately three hours), organisers feel to have fully integrated the public in the deliberative process. According to them, public participants were particularly involved, sometimes with strong controversial arguments. But it is difficult, when looking closer at the minutes, to determine precisely the proportion of time of speech between public and experts. Besides, because of the lack of external evaluation, we cannot establish to what extent public participants feel to have been constructively a part in the deliberative process.

However, the process can be regarded as transparent if we consider that it was made well traceable for participants and outsiders through an extensive presswork — for example, French national radio 'France Culture', one of the partners, transmitted two conferences live — and the use of the internet tool, from communicating the introductory papers to putting the minutes online. Besides, a Nanomonde forum had been opened during the conferences to accompany the process and exchange ideas.[10]

The French Nanomonde can be regarded as a typical example of debate involving experts and public audience. Its strengths are the well-balanced panel of experts from science and research, industry, policymakers, and civil society organisations; the quality of information provided through the website; the high level of transparency of the process with related information; and use of Internet tools. But lack of external evaluation has provided information on the extent the public feels to have been fully involved in the process: the ratio between the number of people involved (120) and experts (6) during the three-hour conference had probably produced inequality between time for public interventions and time for experts' answers and thus might have affected the full involvement of all participants in the process. This first cycle of conferences scores also less good in adopting a generalist approach in the selection of nanotechnology-related themes — the subsequent two ones will be dedicated to more specific issues.

Recently Vivagora refuses to regard recommendations as the main objective or outcome from their deliberative processes: now they claim that these open debates with large participation instead should aim at helping citizens to formulate better questions about science issues rather than trying to formulate recommendations that in short time have to synthesise contradictory positions and interests. This probably means that decision makers behind Nanomonde/Vivagora feel that the original design needed to be developed.

---

[10]www.sciences-et-democratie.net

Chapter 7

Citizens' Conference, Île-de-France

Pål Strandbakken
SIFO Consumption Research Norway, Oslo Metropolitan University,
Postboks 4, St. Olavs plass 0139 Oslo, Norway
pals@oslomet.no

7.1 Introduction

The 'Conference de Citoyens sur le Nanotechnologies' (the Citizens' Conference on the Nanotechnologies) was enacted in order to introduce a more democratic approach to science policy in the Île-de-France region. The conference was organised and financed by the Regional Council. During a four-month project period, 16 citizens participated in a deliberation on nanotechnology and the nano challenge, built around expert lectures and weekend seminars. The end result was perceived as input to the council's science and technology policy in the form of a written statement (see Appendix 1). Consumption and consumer products were not explicitly targeted as topics.

Consumers and Nanotechnology: Deliberative Processes and Methodologies
Edited by Pål Strandbakken, Gerd Scholl, and Eivind Stø
Copyright © 2021 Jenny Stanford Publishing Pte. Ltd.
ISBN 978-981-4877-61-9 (Hardcover), 978-1-003-15985-8 (eBook)
www.jennystanford.com

7.2   Description of the Process

The *purpose* of the Regional Council, as the process initiator, was to generate advice to improve the quality of its decision-making and to more generally initiate upstream communication when approaching an emerging technology. It seems reasonable to assume that the Council regards a broader knowledge and interest base behind decisions as advantageous in itself, but that in addition, the Citizens' Conference might provide regional science policy with a welcome democratic legitimacy. Beyond legitimation, the conference might contribute to increasing public awareness and debate.

The person directly in charge of the conference, Marc Lipinski, was on a six-year mandate for the Green Party and the elected *vice president for higher education, research, scientific, and technological innovation* in the Île-de-France Regional Council. The region is rather large, with as many as 12 million inhabitants, so the region controls and uses large amounts of money for research. The Citizens' Conference was part of an undertaking to develop a politics for science; explicitly defined as an effort to *involve* and *empower* citizens in knowledge production and appropriation. Prior to the conference process itself, the council devoted a year to consultations (2004), and another to the development of an action plan (2005). The action plan defined a set of topics of major interest, so called TIMs, and when that label was granted for a topic, a yearly budget for analysing and dealing with it was guaranteed.

This means that *the initiative* of the process came from a democratically elected public authority and that the purpose was *political*, in a rather wide definition of politics. This also means that the *funding* for the process came from public budgets. Economic compensations for participants were set at €50 per day, in addition to travel costs and accommodation and meals in connection with the weekend seminars. As for the council's total costs, there is some so far unresolved uncertainty. The total sum of €100,000 has been mentioned, but it is not clear if that was for the event itself, for the documentary film made of the event, or both.

From start to closure of the conference, defined as the time the participants were involved, the process lasted for four months, from early autumn 2006 to late January 2007. The main part of the 'effective time' was spent in three training weekends in the months October, November, and December, the last one both in preparation for a public debate with experts and stakeholders, and, after the public meeting, a final session in order to prepare the set of policy recommendations. These were presented at a press conference on January 22, 2007. The effective time was, however, not limited to the training sessions in the weekends, because participants did a fair amount of nano-related homework, read papers, talked with the neighbours, or had family discussions.

The participants were selected as *citizens*, not as stakeholders. Their task was to represent the common voice. Directly responsible for recruiting the group was the the Institut Français d'Opinion Publique (IFOP) survey institute. A mixture of recruitment methods was employed; advertisements, word of mouth, and direct recruitment in the street, etc. Sixteen persons were selected in September 2006. This could never be a 'representative' group in any statistical meaning, but it nevertheless broadly reflected the diversity of the population in the region, with regards to age, gender, and ethnicity. There was, however, clearly an under representation of ethnic minorities and of people with very low education.

The process was organised, defined, and monitored by a Steering Committee, comprising experts in nanotechnology and in participatory democracy. The members of the Steering Committee were selected by Lipinski. In addition he was supported by a coordinating committee, probably made up of Regional Council insiders.

We do not have information of any previous experience with such attempts at using a direct democratic channel into politics on behalf of the Regional Council of Île-de-France, but we suspect that this was an introduction of a new element on the institutional level. However, individuals with relevant competences and experience might have contributed at different points in time.

7.3 Review of the Process

7.3.1 Initiatives and Objectives

To what degree did the process reach its stated goals? At least the delivery of a set of policy recommendations was achieved. In addition, the public debate and the press conferences probably did contribute to raising the awareness of nano as an emerging technology issue in the Île-de-France region. The conference itself ended on a 'positive but cautious' note, but later developments in France indicates that the public reaction to nano has grown more hostile, at least there is a mixed reception: 'The Committee for the Public Debate in France had to suspend the meeting on 1 December in Grenoble, because 100 participants made it impossible for others to make themselves heard. The meeting will be held in another form soon' (Nanoforum Newsletter). These public debates on nano are organised by the mentioned committee (CNDP), at the request of the French government.[1]

We do not know if this national anti-nano activity is in any way connected to the conference in the Île-de-France region, but there is always the possibility that public concern and attention to an issue might trigger fear and uncertainty, regardless of how the conference itself concluded.

7.3.2 Organisation

Three years after the event, the Citizens' Conference could in some ways be regarded as a model for a certain type of deliberative process on nanotechnologies or even for other emerging technologies. When some attempts at democratising science have to do the job in a three-hour session, the Citizens' Conference was able to devote four months. Three plus one training weekends over such a long period also provides time for 'digesting' the rather large amounts of information, for additional learning, and for individual reflection over complex and value-laden questions. This means that the organiser provided time resources for a comparatively satisfactory learning process.

Resource wise, the conference also had almost unlimited access to experts. For the first training weekend experts were selected by the organiser, but responsibility for choosing the type of expertise needed was gradually transmitted to the participants.

7.3.3 Participation

After an initial period of reluctance and suspicion of manipulation and even conspiracies, participants gradually developed more

---

[1]The homepage for this French initiative is http://debatpublic-nano.org/

fruitful approaches to their tasks. And towards the end of the process most of them seemed very involved in the work and were enthusiastic about it. For the conference to successfully end with a set of policy recommendations that represented the views and concerns of the whole group, participants had to commit themselves and to work hard, take part in discussions, and expose themselves. This level of participation was apparently achieved in this process.

The scope and the content of the process seemed clearly defined and explicitly declared to the participants; they were from day one working towards the public debate and the policy recommendations.

We do not have access to material (in the DVD or in other sources of information) that might reveal any interesting alliances or conflicts among the participants. Even if this was not explicitly defined as a 'consensus' conference, it seems clear that a set of unanimous recommendations would have increased its potential future impact. This level of agreement was, however, not achieved.

As mentioned previously, the background of the participants aimed at roughly mirroring the population of the region with regards to gender, age, and ethnicity. This should probably increase the political legitimacy of the event and in addition it was supposed to secure a broader base of interests and views from the group, potentially making its input to the political process more interesting and hence more valuable for policymakers. The number of participants, here16, might be regarded as a bit small; at least when we are concerned with representativity and to have presented a wide range of opinions and ideas. But it is probably easier to achieve internal dynamics and goal-oriented work in a relatively small group as this.

7.3.4 Results of the process

The outcome(s) of the Citizens' Conference sums up to two plus one deliverables:

(1) Chronologically, the first deliverable was the *public debate with stakeholders and experts* in January 2007.

(2) Right after that the *policy recommendations* were fine-tuned and agreed upon and presented at a press conference (see Appendix 1).

(3) In addition, it seems reasonable to include *the film* 'Le Nanos et nous' as a kind of input to the more general debate over nanotechnologies/emerging technologies in the region, in France and in Europe.

The policy recommendations are regarded as the outcomes of the process, and they do actually represent a 'common voice'. The articulated interests of stakeholders were not really present until the public debate, and it does not seem as if they had any decisive influence on the results. This was strictly a deliberative process for citizens and not a stakeholders event. Stakeholders were given the opportunity of influencing the conference outcomes solely on the strength of their arguments. If participants agreed with them, they could choose to present stakeholders' views. The addressees of the outcome are, as we have seen, the directly involved policymakers in the field of education, research, and scientific and technological innovation in the region.

The final policy recommendations document states when it speaks on behalf of all and when there is disagreement. There might have been a gradual movement towards some sort of shared opinion. It is possible that this shared opinion partly results from the formulation of recommendations on a fairly general level. The overall approach of the conference turned out to be positive, but concerned. However, this view was not shared by all. This approach is perhaps a version of the technology optimism stand; we should not try to stop the scientific development and should explore the potential benefits so that we can see a responsible introduction of technology, highlighting environmental issues, health issues, and questions of human rights.

The group decided to focus on five subthemes:

1. Medicine
2. Economy
3. Environment
4. Communication
5. Defence

This was not a thematic 'restriction', but rather a way to organise the subject matter prior to the public debate. When agreeing in these topics, it was easier to agree on which stakeholders to invite to the public debate. The pre-formulated questions for selected politicians and business leaders, and their respective answers or response also provided input to the overall conclusions of the Citizens' Conference.

Most of the political recommendations from the conference tended to fall under the general theme of societal and democratic control with science and technology; often through attempts at institutionalizing common values and the common good:

- Invest more in basic research (proper evaluations of the risks related to nanotechnology is sadly lacking).
- Increase the social awareness of NT (due to our fellow citizens' poor understanding of what is at stake with the development of nanotechnology, the possibility of misuse cannot be ruled out).
- Develop toxicology and 'societal' projects.
- Invite NGOs to participate in the problematic formulation of the call for projects.
- Transfer the recommendations from the Citizens' Conference to national and European levels.
- The most important general recommendation was, however, the call for an independent body consisting of high-profile politicians, scientists, philosophers, members of ethics committees, and representative citizens (mainly from NGOs) to ensure respect for ethics, to oversee laboratory research and further research and funding, establish traceability, make lists of hazardous products, disseminate findings and conclusions, etc.; all funded by the Regional Council.

And then they added a very interesting wish: 'we would like our group to be kept fully *informed of the extent to which its recommendations are taken into account'* (emphasis in the original).

Some of the other recommendations aimed at some kind of generalisation of the process they had been a part of, to take steps to share the experience. The idea was to initiate follow-up debates on national and European levels.

As for questions of what happened after the event, after the press conference on January 22, 2007, we will return to that under the section Summarising Appraisal of this chapter (Section 7.5). There we will also comment on some of the more specific recommendations that came out of the event.

7.4 Deliberation Criteria

From Cohen's (1989) perspective, there was an agenda and a theme. The *agenda* was to provide input into regional science policy, and the *theme* was nanotechnology/nanoscience. But it is not obvious that the existence of a theme should be regarded as a 'thematic restriction'. The definition of a theme is rather a precondition for having a fruitful deliberation. Beyond that, it is hard to recognise any constraints on the process. One might say, however, that participants, representing the population of the Île-de-France region, were led into a logic of weighing potential risks against potential benefits of an emerging technology (Beck, 1992). It would perhaps be possible to read this as shrewd manipulation on behalf of political authorities, but with such a paranoid perspective democracy is hardly possible at all. In the defined context, we chose to regard the Citizens' Conference as a *free discourse.*

We do not have extensive knowledge of the forms of interaction between the participants, but it seems as if it started out as a mix between an educational seminar and a focus group that organised itself in order to deal with the challenge of addressing the relevant themes. It is fair to call it a *reasoned process.*

Scientific and other reasoned arguments were brought into the process on the participants' initiative. After the first training session, where the organiser provided expert lecturers, the group itself was responsible. For the whole process, 17 lectures were held by experts, covering disciplines like physics, toxicology, biology, and ecology, in addition to anthropology, economics, science, history, and law. Thus, participants were exposed to a variety of opinions through the process, from external experts and from other citizens. The whole process was designed for mutual learning, and no one in the group entered with a fixed set of positions and opinions on nano.

On the other hand, 16 citizens entered into such a process with their political and ethical values and perspectives, their worldviews, and their more specific attitudes to science and technology, like being mainly technology optimists or the contrary. Then again, this is probably what makes deliberative processes of this kind interesting.

Regarding transparency, the self-organisation of the group, as well as its lack of formal leadership, combined with the set of tasks that were known by all, seem to have made this Citizens' Conference process transparent as an event, even if its formal status in the political and administrative landscape was not necessarily clear.

To what extent were participants equal? It is not uncommon that educated middle-class persons tend to take or be given leading roles in these kinds of mixed groups (juries, focus groups, etc.). This probably has much to do with self-confidence when it comes to writing and formulating points of view and familiarity with the setting where you make decisions in meetings and the like. It seems, however, as if the participants in the Citizens' Conference to some extent managed to rotate the more or less leading roles. It is difficult to decide after the event whether a flat structure was maintained through the process, but it is also open to debate if this maintenance is necessary. Why should it be a problem if the group developed some informal leadership structures? As long as it was not dominated by one or two strong manipulators, such structures do not have to be problematic. In Cohen's criteria the main point is that participants are recruited as equals and no participant enters the process as more important than any other. What happens in the process should be seen as another matter.

The logic behind these kinds of deliberations usually is to arrive at a certain degree of consensus; indeed citizens' deliberations are often called 'consensus conferences'. We do not know how and when votes were taken during the process, but there had been some voting activity, since the policy recommendations states it be undertaken when there were disagreements. If consensus was aimed at, it was not reached. In Section IV, 'Our opinion', of the policy recommendations (Appendix 1), the text points to differences of opinion on a number of themes, such as ethics, but also on the general perspective on technology (*most* of our group are in favour of nanotechnologies for a large number of reasons). If a consensus driven process was the ideal, it was not completely reached in the Citizens' Conference.

7.5 Summarising Appraisal

The way the Citizens' Conference was organised and executed makes it reasonable to regard it as a typical deliberative process. The Regional Council supplied the resources in the form of locations, covered travelling costs, etc., and paid for/supplied intellectual resources in the form of lectures chosen by the group. Under the rather broad frame of 'nano', participants defined their roles and their themes and organised their work.

The strengths and weaknesses of the process should perhaps be analysed in comparison with similar processes in other places/cultures. Some things seem rather clear, though: on the positive side here we once more have to point to the amount of time and resources devoted to the task. What we in retrospect might see as weaknesses are mainly two factors. First, the mandate tended to get too wide. With nanotechnology and nanoscience as an enabling technology influencing a whole range of processes and technologies, the idea of saying something interesting about the phenomenon might appear too ambitious, even with the aforementioned resources. Second, the links between the conference/the deliberation and the policy-making level were a bit unclear; they depended very much on the enthusiasm of the initiator and on his ability to translate policy recommendations into research practice.

The first point could very well prove to be relevant for more or less all deliberative processes on nano of this 'generation'. Nanoscience and applications are presented as giving more or less unlimited possibilities while potential negative effects remain uncertain. A body of citizens then easily tends to give rather unspecific comments and recommendations. A question that we deal with elsewhere is one about convergence. If more or less 'all' citizens' deliberations on nano conclude similarly, regardless of time use and other resources, such general deliberations might have exhausted their potential.

The question of the connections between the exercise and actual policymaking is also crucial. In the Île-de-France conference, the Regional Council for higher education, research, and scientific and technological innovation initiated the event in order to broaden their views on an emerging technology. The citizens' influence on research policy was potentially direct. As for suggestions like 'invest more in basic research' and 'develop toxicology and 'societal' projects', they are rather easy to consider and even implement. One suspects, however, that this is more or less what the council would have done anyway. To include NGOs to participate in development of project calls is slightly more original.

The more specific recommendations, which are generally thought of as being easier to implement, might actually be trickier, however. The Citizens' Conference did recommend a labelling scheme for nanoproducts, but this was not supposed to be possible to implement on the regional level. A scheme for establishing a kind of cross-disciplinary nano observatory was rated as interesting, but it should probably rather be established on a national level or on the European level (such an institution has actually been debated on an EU level).

Lipinski's post-event comment was that most of the recommendations could not effectively be dealt with by the Regional Council, mainly because they were too general. National defence and national economy also fall outside the regions' jurisdiction. Even 'precise and clear labelling' is difficult to organise and enforce on a Regional level. This means that in order to use deliberative processes on an emerging technology to achieve some kind of democratisation of science, one would have to initiate follow-up debates on national, European, and international levels.

What can we learn from this process? Lipinski was two years later asked what, if anything, he would do differently, if he had had another go (Lipinsky, November 2009). Most of the afterthoughts were concerned with how the deliberate process worked within a political system that has a science policy that is limited by the national and European levels. The potential for following-up should have been considered more thoroughly in advance.

On a general level, the French nano unrest after the Île-de-France conference and Nanomonde is interesting. It seems to confirm the observation that more information does not necessarily breed public acceptance.

References

Beck U (1992) *Risk society. Towards a new modernity.* London: Sage Publications.

Cohen J (1989) Deliberation and democratic legitimacy, in *The Good Polity* (Hamlin A, Pettit P, eds), Oxford, Basil Blackwell.

CNDP. French debates on nano. Available at http://debatpublic-nano.org/index.html

Hover D (Director) (2007) *Le nanos et nous* [DVD]. France: Riff International Productions

La Région Île-de-France (2007) Policy recommendations from the Conference de citoyens sur les nanotechnologies. Available at http://espaceprojets.iledefrance.fr/jahia/Jahia/bca/NanoCitoyens/site/projets

Nanoforum (European Nanotechnology Gateway). *Nanoforum Newsletter.* Available at http://www.nanoforum.org

In addition, we had a seminar on *The future of deliberative processes in emerging technologies* at the Danish Board of Technology in November 2009, where, among others, Lipinski was present, and answered questions on the Citizens' Conference.

Appendix 1

Policy Recommendations from the Citizens' Conference

Introduction

As citizens of Île-de-France, with all our differences, distinctive characteristics and diversity, we have come together to debate the key issues surrounding the development of nanotechnologies.

We are addressing our opinions and recommendations to the elected representatives of the regional council.

The world of nanotechnology excites our curiosity while raising a myriad of questions. The field of nanotechnology is highly complex and difficult to grasp with the result that merely accessing information presents a challenge in itself.

The sheer complexity of the nanoworld gives us considerable food for thought.

This complexity is causing growing concern. Firstly, because there is a risk that those who possess knowledge end up taking decisions on our behalf without keeping us informed, which represents a genuine threat to democracy. Moreover, due to our fellow citizens' poor understanding of what is at stake with the development of nanotechnology, the possibility of misuse cannot be ruled out.

It strikes us that within this complex universe, the sheer difficulty in grasping the problem set and taking the necessary decisions means that we have an obligation to take the time needed to enhance our understanding. Major efforts are required to fully participate in the debate.

I. Observations by field

1. Evaluation of risks

In our view, proper evaluation of the risks related to nanotechnology is sadly lacking. This inadequate response primarily stems from the fact that the use of nanotechnologies is recent and the tools needed for an effective evaluation is not at our disposal. There are those within our group who wonder whether we might be witnessing a deliberative strategy on the part of diverse players (scientists, manufacturers, politicians) top obstruct this process of evaluation or at least conceal its findings when it does take place.

This absence of evaluation compounded by a lack of understanding on the part of the general public as to the risks leaves the door wide open to all kinds of presumptions and concerns which are exacerbated by fears over past controversies surrounding issues such as asbestos, nuclear energy, etc.

However, we are well aware that an excess of evaluation and an overly broad application of the principle of precaution may put a brake on research and undermine our position in relation to international competition with the attendant risk of negative economic consequences.

We believe that risks can be taken in certain fields such as healthcare or medicine, but not in others.

All the players and experts whom we have met recommended in increase in credits and investments to support the evaluation of risks related to the use of nanotechnology.

2. Ethics

For us, the notion of ethics means acting with respect for human beings and our environment. We view it as an essential protective barrier to prevent any misuse and this is vital for our survival.

Ethics raises the question of how the individual relates to the society at large. We do not all have the same understanding of ethics within the group, while remaining fully aware of this necessity.

It can be seen that ethics is present (in a direct or underlying manner) in the message coming from all players we met. It would seem that the responsibility of the various players including that of consumer-citizens lies at the very heart of ethics.

3. Information and communication

Information on nanotechnologies is elitist and restricted to specialists (complex articles in magazines which are often impenetrable to the layman).

The general public is badly informed. Why?

- Novelty of the subject
- Inappropriate information
- Lack of awareness and reticence over finding out more
- Disinterest etc.

We ourselves knew little about the subject before embarking on our training in the context of the citizens' conference.

Now that we have found out more, we have become more conscious of our role as citizens. We have developed an interest in a subject which has awakened our curiosity and are generally more aware and pick up on articles in newspapers and magazines much more often.

Knowledge of the subject is therefore a prerequisite to getting involved.

4. Legislation and regulation

We note that there are no regulations specific to nanotechnologies. They fall within the scope of regulations concerning other products (chemicals, medicinal products, etc.). For instance, nanotechnology was not included in the REACH directive, despite pressure from associations. It would seem that the public authorities are not favourably disposed towards legislating on this subject for the time being.

In our view, this absence of regulation could lead to misuse and therefore an increased risk of potentially hazardous products being placed on the market. There is a lack of responsibility which leaves the consumer with no avenue for recourse in the event of loss.

II. The sectors of application

1. Medicine and healthcare

This is the most tightly regulated sector, in which risks are most effectively controlled and at all levels: control of raw materials, handling of products, testing and marketing authorisations for medicinal products, etc.

This is also the field which communicates most effectively by means of a broad dissemination of information aimed at the public. This information concerns the discovery of therapies for the treatment of debilitating or deadly diseases (e.g., SFDs, cancers, etc.).

It is also the sector where the liability of practitioners is the most exposed hence the possibility of recourse for victims.

The contribution on nanotechnology is particularly promising in the field of medicine and healthcare (diagnosis and testing, implants, DNA chips, etc.).

Secondly, the prevention of occupational illnesses is a field which is also now starting to benefit from advances in in nanotechnology (patches, sensors, filters, detectors). But these advances also lead to potential misuse and risks such as attempts to boost physical and intellectual performances, genetic engineering or the psychological control of individuals.

2. **Military aspects**

   The arms industry has taken considerable strides forward thanks to use of nanotechnology in all fields (aeronautics, naval and land forces).

   The FELIN programme, the first French military programme for nanotechnology equipment, aims at making the field soldier as efficient as possible.

   This field continues to be protected by the notion of defence secrets hence the difficulty in obtaining information. However, the military application of these advances has important consequences for civil applications (e.g., GPS, mobile telephone, etc.).

3. **Information and communication**

   We carefully took note of the message delivered by the CNIL (French data protection authority) and the position it has adopted with regard to nanotechnologies. These enable technical advances in the storage of personal information (thanks to RFID chips) as well as their dissemination. The CNIL recommends the deactivation of these means of communication (e.g., after purchasing) to avoid data being subsequently communicated.

These new technologies do not require fundamental changes to the 1978 Data Protection Act but the increase in computerised data raises problems as to the applications of these rules and the means implemented.

We note a lack of regulation as to coding of chips and a resulting risk of unauthorised access to and/or the leaking of such data.

4. Environment

We note that considerable progress can be achieved in terms of the quality of water, air, and land. We have become aware that industries active in the environmental field are able to achieve significant improvements in technical performances while slashing operating costs.

Moreover, it would appear that thanks to these new technologies, savings in natural resources are achievable along with a significant reduction in energy consumption (e.g., depollution of water, purification of air, treatment of waste).

However, the use of nanotechnology also poses risks for the environment: contamination of the food chain, problem of recycling 'nano' waste as well as pollution of tap water.

As things stand, we note that there has been no response to the risks posed to the ecosystem and the environment by dissemination of nanoparticles.

5. Economy

Enormous sums are at stake. Some countries have fully committed to the development of nanotechnologies and France could hardly be described as leading the field.

The granting of subsidies by the Île-de-France Regional Council is not tied to the creation of jobs.

No official study has been conducted into the creation of jobs and businesses within the region but we note its determination to play a more active role in supporting research and training.

We have had no precise response as to reductions in production costs related to nanotechnologies.

Nanotechnology is already proving its worth in the field of micro-electronics (STMicroelectronics has informed us that 80% of its products contain nanostructured components).

As for delocalisation, it would seem that skill centres will remain in France but there is no guarantee that production facilities will do likewise.

III. Our definition of the problem set

Participative democracy is considered as essential and politicians are keen to establish how citizens will react to these questions. The region is keen to obtain the assent of its citizens, especially since the development of nanotechnologies directly concerns them.

With this in mind, the Regional Council and therefore our group are faced with the following questions:

- Considering the strong position and major assets of the Region on both national and European levels, what is the right policy to adopt?
- Should we develop nanotechnologies (setting up a 'nano development' hub)?
- What part of the budget devoted to research and business creation should the Regional Council assign to the nanotechnology sector?

IV. Our opinion

Most of our group are in favour of nanotechnologies for a large number if reasons. Nanotechnologies clearly represent an opportunity for progress and even hope for the world today and tomorrow whether in the fields of healthcare, daily life, our environment, and our surroundings. These nanotechnologies open up the possibility of improved aid to developing countries.

In addition, nanotechnologies are indispensable from an economic viewpoint. Their development has become key to the creation of wealth and jobs.

However, we feel it important to set up conditions:

- We are opposed to a 'Big Brother' society
- It would be unacceptable for the economic gains from nanotechnology to be achieved at the cost of ethics

- We are keen to see the establishment of rules governing the development of nanotechnology since nanoparticles are potentially hazardous and the risks to the environment and life forms are real. We feel responsible for our planet and our surroundings. This implies an obligation to respect the environment and the ecosystem.

V. Our recommendations

- Each manufacturer must accept moral responsibility for the ecological and health risks to which the development of nanotechnologies exposes us
- We are calling for specific measures concerning manufacturers in the Île-de-France region: taking precautions, setting up a protocol for the handling of products containing nanoscale structure. The Region will draw up a transparency charter which they should apply: labelling, evaluation of risks, etc;
- It can be seen that nanoproducts have already found their way onto the market despite inadequate research into their hazards. Current health and environmental legislation is poorly adapted to the use of nanoscale materials. Given this inadequacy, it strikes us as vital to respect the principle of precaution
- Precise and clear labelling must be affixed to products originating from nanotechnology in order to keep consumers fully informed.
- With regard to public information on nanotechnology, we would like to see an extensive campaign using terms accessible to all and across all media (press radio, TV, Internet, etc.).
- We would also like to see greater budgetary resources assigned to the CNIL. We expect to see the setting up of actions to build awareness of respect for personal liberties at the EU level.
- We recommend a partnership with the main established consumer associations which can serve as an interface with citizens.
- We are keen to see the strengthening of research, the key element for exploring the challenges relating to the infinitely small. This research effort should be focused

on clear scientific objectives; indeed, nanoparticles no longer respond to the traditional laws of physics and require expertise in a highly specific discipline with its own exploratory tools.

In conclusion, to represent the interests of citizens we would like to see the creation of an independent body, comprised of:

- high-profile politicians
- scientists
- philosophers
- members of ethics committees
- representative citizens (members of recognised associations)

This body's remit will be to:

- ensure respect for ethics;
- oversee laboratory research;
- deliver an opinion on the continuation of this research effort;
- verify that the funds invested by the Region are being put to proper use;
- draw up a list of potentially hazardous products;
- establish traceability of nanoparticles from production through destruction or rrecycling;
- disseminate its findings and conclusions to citizens;
- etc.

This Regional Council will provide this body with the financial resources needed to fulfil its remit

Finally, we would like our group to be kept **fully informed of the extent to which its recommendations are taken into account.**

Chapter 8

Nanotechnology Citizens' Conference in Madison, USA

M. Atilla Öner[a] and Özcan Sarıtaş[b]

[a]*Management Application Research Center, Yeditepe University, 26 August Campus, IIBF 648, 34755 Ataşehir, İstanbul, Turkey*
[b]*Manchester Institute of Innovation Research, University of Manchester, Harold Hankins Building, Manchester, M13 9PL, UK*
maoner@yeditepe.edu.tr

8.1 Introduction

The Center for Democracy in Action[1] (CDA) organises consensus conferences involving small groups of citizens who go through a learning process on a given technological issue, engage experts, and develop an assessment of the key issues they identify as critical (Einsiedel & Eastlick, 2000). In line with the overall purpose of examining the deliberative processes in nanotechnology development, we will have a close look at the nanotechnology consensus conference in Madison (USA), which was held in April

---

[1]http://cdaction.org

Consumers and Nanotechnology: Deliberative Processes and Methodologies
Edited by Pål Strandbakken, Gerd Scholl, and Eivind Stø
Copyright © 2021 Jenny Stanford Publishing Pte. Ltd.
ISBN 978-981-4877-61-9 (Hardcover), 978-1-003-15985-8 (eBook)
www.jennystanford.com

2005 in three sessions. Our review is mainly based on the academic literature and online materials available.

The consensus conference on nanotechnologies was hosted and funded by the University of Wisconsin's Integrated Liberal Studies Program and the Nanoscale Science and Engineering Center, an NSF-funded program at the University of Wisconsin.[2]

8.2   Description of the Process

Thirteen Madison-area citizens were involved in the three sessions of the conference which were held on April 3, 17, and 24, 2005. Each session was approximately four hours long. During this time, the participants studied nanotechnology, met with a group of interdisciplinary experts, and developed a final report. A final press conference was held on April 28, 2005, with the participation of elected officials and the media, where the final report was distributed. Copies of the report were also sent to all Wisconsin legislators. The Citizens Coalition formed out of the consensus conference and its development was relatively organic. We will describe the overall process in detail in the following sections.

The overall process of the consensus conference was initiated by the CDA. CDA is an organisation dedicated to increasing the spaces, places, and events for inclusive dialogue and strategic grassroots action.

This conference was CDA's first experience with deliberative processes on nanotechnology. Collaborative partners in this process were Daniel Kleinman, University of Wisconsin–Madison; Maria Powell, UW-Madison post-doctoral student; UW-Madison Nanoscale Center; UW-Madison Integrated Liberal Studies Program.

The process resembled a 'process aiming at the representation of the common voice' (Einsiedel & Eastlick, 2000).

Adequate funding was crucial for the organisation and implementation of the conference to cover staff time, outreach materials, expert and participant stipends (if offered), food, child care, transportation, and a variety of other key elements of the process. Although many of these were mundane logistical matters, they could have significant impacts on outreach, the diversity of the participants, and the quality of deliberation.

---

[2]http://cdaction.org/nanotechnology_citizen_conference

In the consensus conference, the project team faced some fairly significant funding limitations in carrying off the initiative. Early European consensus conferences were undertaken with budgets between $100,000 and $200,000. The low-budget 1997 Boston consensus conference on telecommunications was run for $60,000. These efforts relied on numerous paid staff members, including organisers and administrative support personnel.

By contrast, the multiday forum was carried out with a budget of approximately US$ 6000 and four part-time organisers. One of the organisers was a full-time professor, and another was a half-time post-doctoral research associate, and both were paid out of existing university salaries. Neither, however, had reduced standard workloads to create space to undertake this initiative. The remaining two organisers, leaders of a community-based organisation with experience in hosting dialogue-centred events, were paid very small stipends as facilitators and they worked beyond their paid time.

The consensus conference was also integrated into a liberal studies undergraduate seminar taught by the project leader at the University of Wisconsin–Madison. Students in the class were involved in every aspect of organising our forum. One might conclude that the person-hours gained by this manner of operating amounted to a substantial in-kind contribution of labour, which saved a substantial amount of money. However, the project team failed to think clearly through all of the ways to integrate students into the conference organising process, and at times, they probably sent mixed signals about the students' responsibilities. Consequently, though it is true that the students were actively involved in the process, the core project team did not utilise their skills and energy as effectively as they could have.

The amount of time allotted for a conference and for preparation can have important impacts on nearly all aspects of the forum. Typical consensus conferences have taken between 10 and 18 months to organise and carry off. The project team spent less than 4 months from recruitment to press conference. Their abbreviated time frame was required, in part, because the consensus conference was integrated into an undergraduate seminar. This short time frame was a disadvantage in some ways. More time from the preparation phase to the final event would have allowed the project team to do much more comprehensive outreach and, more importantly, improve the deliberative process substantially by allowing more time for interaction among participants and for interaction between participants and experts.

Determining conference length has clear trade-offs. The nanotechnology consensus conference involved three 3–4-hr sessions plus a press conference. Holding more than three sessions and allowing more time for each of them might have permitted citizens to examine the issues at stake more thoroughly. However, asking citizens to commit to participate in a longer conference would have created significant barriers for people with children, people who have little flexibility with their time because of work commitments, and others who might not be able to spend that amount of time for various reasons. In this connection, a longer consensus conference might very well have lowered the social diversity of the participants in the conference.

8.3 Review of the Process

8.3.1 Initiatives and Objectives

Overall, it can be said that the process reached its internal goals by involving and educating 13 citizens on the possible risks involved in the use of nanotechnology and preparing a final report which was sent to Wisconsin legislators.

However, an Internet search in 2007 revealed that there was a complaint about the legislators for not having used the recommendations of the report.

8.3.2 Organisation

We can conclude that the project must have had reasonable amount of resources as it was completed successfully. The faculty members of the University of Wisconsin and other experts involved in the projects were competent individuals in their respective fields. We may view the process as an outsourcing project for CDA.

There are several reasons the conference was carried off successfully in spite of substantial funding limitations, and that these issues are important for organisers to consider when picking staff to organise consensus conferences (Kleinman *et al.*, 2007). It is particularly important to choose at least some staff that has community-organising experience. All four staff involved in the conference had years of experience with participatory mechanisms in a variety of community settings. Consequently, they had knowledge that prepared them to facilitate many of the practical aspects of organising the conference smoothly and efficiently without much administrative support.

Perhaps more important, several of the lead organisers had connections in the Madison community that were very helpful in the outreach efforts, particularly with audiences and groups that might be hard to reach through mainstream channels (e.g., minorities, low-income people, less-educated citizens). Most of the staff had experience in facilitating deliberative group work and/or consensus processes with diverse groups and they understood the need for good facilitation throughout these processes. Several of the staff already knew each other well, and all four staff members spent considerable time with each other before and during the conference, planning the process and building trust. This made the process itself and communication throughout it run much more smoothly than it likely would have if the staff had not taken the time to plan together, build relationships, and get to know each other's working styles.

On the basis of their experience, the core project team proposed (Kleinman *et al.*, 2007) that in addition to having adequate funding to hire enough staff, a conference is more likely to meet the intermediate goals if the staff has some organising and facilitation experience and has strong relationships with each other and with the community. Ideally, future conferences would have a more diverse staff and expert panellist group than this conference in which all of the core staff were Caucasian and only one of the expert panellists was non-Caucasian.

Given that many consensus conferences are organised within academic or government institutions, finding diverse staff with organising experience might be challenging because typically academics and government employees are not trained in community organising, and minorities are significantly underrepresented in these institutions. However, some academic departments (e.g., sociology, community education, and extension) may include professors and graduate students with organising and facilitation experience, and consensus conference organisers would be well advised to tap into this expertise. Moreover, diverse individuals with organising backgrounds, from community and neighbourhood groups, could be recruited for consensus conference projects, even if they are not within academic settings. These individuals not only bring valuable knowledge, perspectives, and experience to the process, they also are likely to have established relationships with people in the community that will greatly assist in outreach and trust building throughout the process.

8.3.3 Participation

The participation of 13 individuals from the area may be adequate to discuss the answers to the necessary questions through discussions with experts, but was definitely not adequate for a democratic involvement project based on a deliberative process.

The Madison Citizens' Consensus Conference took place over three weekends at the University of Wisconsin student union, an oft-remodelled structure opened in 1928, in Spring 2005 (Powell & Kleinman, 2008, Kleinman *et al.*, 2011).

Calls for the conference took place for approximately two months before the conference and included press releases to all the major newspapers, television and radio stations, and local papers. In addition, the initiative was announced on several community websites, and along with a number of students, several community groups were visited describing the consensus conference. In the end, 18 applications were received and 13 participants were selected. One participant who had previously participated in a consensus conference and two participants who had substantial knowledge about nanotechnology were not accepted.

The group of panellists was demographically diverse, included younger and older members, women and men, and people from different races and religions and an array of occupations. The citizen group was diverse by several measures. The panel of 13 members included six women and seven men. Ten of the participants had incomes at or below US$47000, and three had incomes higher than this. Their ages ranged from 19 years to late 60s, with a fairly even distribution across decades. Ten of the participants were Caucasian, and three were people of colour (Iranian, Latina, and African American). Two of the participants had high school degrees, two were working on associates' degrees, seven had college degrees, and two had master's degrees. A wide array of occupations was represented. Applicants included single individuals, members of couples, young parents, and retirees. Participants were affiliated with several different religious denominations. Despite being a relatively diverse group, it obviously did not fully reflect the demographic profile of the Madison, Wisconsin area.

After receiving commitments to participate from those applicants, who were believed to be best contributors to a well-rounded citizen panel, several relatively short readings from a range of sources were sent to panellists, including articles from *Science* magazine, the Royal Academy of Sciences, an academic journal, the U.S. Department of Agriculture, and the Erosion, Technology, and Concentration (ETC) Group. Since nanotechnology has just been coming to prominence and is likely to affect many areas of economic and social life, readings were included that covered issues ranging from the prospective medical and agricultural uses of nanotechnology to the military devices that might be created by the nanoscale tools and the possible environmental impacts of nanomaterials. In addition to the documents sent to participants, they were given a list of websites that provided further information on the topic.

In the first session of the consensus conference, citizens were asked to generate a set of questions to be considered by 'experts' at the second session. While this was the goal, panellists were instructed to begin the first session by simply discussing the materials they read. Panellists were broken into three small groups, and two professional facilitators circulated among the groups to aid in the flow of discussion. These small group discussions were wide ranging, sometimes considering issues such as the group's trust of the government and other times narrowing in on topics such as appropriate principles for regulating nanotechnology development.

After meeting in small groups, the entire panel came back together and was led in their further discussions by the facilitators. Panellists were asked first to list categories of topics and then specific questions. Topics fell into eight areas with a fair bit of overlap between them. The areas were 'ethics and theology', 'human health and the environment', and 'citizen participation and input'. The questions that citizens were interested in asking the expert panel ran from very broad and philosophical (e.g., 'Should we cure every disease because we can?') to very concrete and focused ('What specific nanotechnology research is being done at the University of Wisconsin? How is this research being funded?').

Organisers took the set of questions the facilitators had written down on an easel in front of the group and attempted to organise it by category and to eliminate repetition of questions. The draft list was circulated to panellists between sessions and the project team prepared a clean list in advance of the citizen–expert meeting.

The second session was open to the general public. It received some advance publicity and, in addition to the citizen and expert panellists, there were some 30 audience members in attendance. The citizen panellists sat at the front of the room with the expert panellists at their sides. The audience was in rows in the back of the room. This session was orchestrated by two facilitators. Citizens asked the questions they had generated at the previous citizen session.

The expert panellists — who included, among others, a chemist and an engineer doing research on nanoscale materials, a public policy analyst, and an ethicist — were permitted to respond to any question they wished. Citizens took notes during the meeting, and this session was covered by the local television and radio.

In the third consensus conference session, the panel again broke down into subgroups and discussed the array of topics they wanted to consider in the final report and the positions they wanted to take. Notes taken by student organisers at these smaller sessions were integrated into a single document that was projected onto a screen for all members of the panel to see. Then, in a facilitated discussion, the group attempted to collectively craft the final report. Amid substantive discussion, there was also consideration of procedural matters. Since the meaning and implications of formal consensus processes were never explicitly elaborated, consideration was given to what the group would do if there were disagreement among panellists. Informally, the group agreed that they would try to include positions in the report that all members could accept, but they did consider the possibility of issuing 'minority positions'. Also in addition to substantive discussion, panellists considered strategy. Thus, for example, they considered how the framing of specific topics might affect the response of public officials.

At the end of the third session, the group had produced a host of recommendations. The project team did some editing to the language of these recommendations and drafted a brief introduction to the report. They then circulated the document among the panellists before finalising it.

8.3.4 Results of the Process

The process resulted in 13 informed citizens and a final report was sent to Wisconsin policymakers (Appendix 2).

The report's recommendations fell into six categories including 'health and safety regulations', 'media coverage and information availability', and 'creation of government bodies', 'research and research funding', 'military surveillance', and 'public involvement'. The specific recommendations were as follows:

1. Development of a 'clear and precise definition of nanotechnology' for regulatory purposes
2. Effective mechanisms to be provided by the government at the local, state, and federal levels for citizen involvement in nanotechnology policy development (e.g., sponsoring citizen fora)
3. Guarantee of public access to 'the results of nanomaterial safety and toxicity tests done by private corporations'
4. Nanotechnology not be used to develop weaponry

The final report was released at a press conference at the Wisconsin state capitol. There were some half a dozen state legislators as well as members of the press, non-profit organisations, and civil servants in attendance. Several of the citizens made opening statements, and then each of the elected officials was given an opportunity to speak.

Across the month from the first session through the press conference, Madison's consensus conference on nanotechnology was widely covered on radio and television and in newspapers. The coverage continued into the spring and summer.

8.4 Deliberation Criteria

A deliberative process is a procedure where 'citizens can propose issues on certain topics for the political agenda and participate in debate about those issues' (Cohen, 1997). Cohen (1989) identifies four criteria for an ideal deliberation: It is a free discourse; participants regard themselves as bound solely by the results and preconditions of the deliberation process, it is reasoned; parties are required to state reasons for proposals, participants in the deliberative process are equal; and the deliberation aims at rational motivated consensus.

This section assesses Madison's consensus conference based on these criteria.

1. In the Madison's consensus conference, the citizens were entitled to a say in all matters on nanotechnologies, which would potentially affect their lives. They were able to state their ideas on the topic based on the background information (e.g., reading materials) they were given and to offer insights that experts would not otherwise consider.

2. It was a reasoned process as the citizen panellists drafted recommendations at the end of the process on the basis of their reading and discussion sessions (one on their own and one with a panel of experts). Each recommendation was explained and justified in detail with the underlying rationales and assumptions.

3. Thirteen Madison area citizens were involved in the process. These citizens from a variety of backgrounds were gathered to participate in an innovative democratic process for obtaining lay perspectives on the future development of nanotechnologies. In this respect they were all treated equally in the process and were encouraged to think collectively about nanotechnologies to discuss their uncertain behavioural characteristics.

4. As referred to in its name, the deliberative process of the Madison's conference was consensus-oriented. During deliberation, a group of citizens was recruited and given background materials on the topic. The citizens then worked together through a facilitation team in order to prepare questions to be asked to the 'content experts'. On the basis of the background materials and discussions with experts the participants produced a list of recommendations based on the consensus achieved throughout the process.

Having assessed the Madison's consensus conference briefly, we can conclude that the process meets Cohen's four criteria for ideal deliberation. Besides the process of the deliberation, the follow-up activities carried out after the process can be considered as a proof of a good example for a consensus conference. These will be summarised in the following section along with some of the shortcomings of the process.

8.5   Summarising Appraisal

The conference can also be considered as a good example of a deliberative process as the efforts led to the presentation of final recommendations to the state-level politicians. The panel members subsequently created a 'citizen's coalition' on nanotechnology. They launched a 'science cafes' programme and continued working on nanotechnologies. A website was also set up in which the group has published essays on the issues related to the governance of nanotechnologies. The citizen's coalition has later been renamed Nanotechnology Citizen Engagement Organisation — nanoCEO. The need for dialogue with civil society organisations have been emphasised since then.

As discussed above, the Madison's consensus conference carried most of the characteristics of an ideal deliberative process. However, it also had certain limitations. According to Kleinman *et al.* (2007), the Madison conference had no direct impact on policy. There was a relatively diverse panel, which conducted a successful deliberation and overall felt empowered at the end of the process. Kleinman *et al.* (2007) also mention some other shortcomings of the process, such as the lack of a template at the beginning of the work. However, throughout the process they created a useful toolkit for future consensus conference organisers by keeping good records of their meetings and having detailed discussions of the factors they considered during the design and implementation of the programme.

Overall, the Madison experience proved that organising consensus conference with the aim of empowering the citizens and engaging them in political action would allow less powerful social groups, which is typically the majority of the society, to have control over technological development and assessment. Laurent (2009) relates this notion of inclusive consensus conference with Sclove's 'technological pluralism', which advocates the involvement of social groups, not necessarily experts, in knowledge production. In this respect, the consensus conferences grant power to lay citizens and involve them in the decision-making by enabling society to have control over technological choices.

The case of Madison can therefore be seen as an important exercise in terms of building citizens' perceptions of their capacities to consider scientific and technological issues and institutions. Powell and Kleinman (2008) report that a majority of the citizens involved in the process felt they gained knowledge and efficacy on nanotechnologies, and some of the participant citizens continued working with the researchers in Madison. It will be through the inclusive and consensus-oriented processes that citizens will be involved in decision-making processes and will in turn shape the policies and practice participative democracy.

References

Center for Democracy in Action. Available at http://cdaction.org/ nanotechnology_citizen_conference.html

Cohen J (1997) Deliberation and democratic legitimacy, in *Deliberative Democracy: Essays on Reason and Politics* (Bohman J, Rehg W, eds), Cambridge, MIT Press, MA, 67–91.

Cohen J (1989) Deliberation and democratic legitimacy, in *The Good Polity* (Hamlin A, Petit P, eds), Oxford, Basil Blackwell.

Comp NJ. Nanotech: Small Wonder. *The Daily Page*, October 8, 2012. Available at http://www.thedailypage.com/isthmus/article.php?article=7802

Einsiedel EF, Eastlick DL (2000) Consensus conferences as deliberative democracy: A communications perspective, *Sci Commun*, 21, 323–343.

Kleinman DL, Powell M, Grice J, Adrian J, Lobes C (2007) A toolkit for democratizing science and technology policy: The practical mechanics of organizing a consensus conference, *B Sci Technol Soc*, 27(2), 154–169.

Kleinman, DL, Delborne JA, Anderson AA (2011) Engaging citizens: The high cost of citizen participation in high technology, *Public Underst Sci*, 20 (2), 221-240.

Laurent B (2009) Replicating participatory devices: The consensus conference confronts nanotechnology, CSI working paper series, No. 018. Centre de Sociologie de l'Innovation. Available at http://www. csi.ensmp.fr/Items/WorkingPapers/Download/DLWP.php?wp=WP_ CSI_018.pdf

Nanowerk (2007) Nanotechnology risk discussion — Where is the public? Available at http://www.nanowerk.com/news/newsid=1227.php

Powell M, Kleinman DL (2008) Building citizen capacities for participation in nanotechnology decision-making: The democratic virtues of the consensus conference model, *Public Underst Sci*, 17, 329–348.

Appendix 2

Written Submission from the Citizens' Coalition on Nanotechnology

To the National Nanotechnology Coordination Office on behalf of the Nanoscale Science, Engineering, and Technology (NSET) Subcommittee of the Committee on Technology, National Science and Technology Council (NSTC)

January 31st, 2007

Cate Alexander Brennan
Communications Director
National Nanotechnology Coordination Office
Arlington, VA 22230

Dear Ms. Brennan:

In April 2005, a group of ordinary citizens in the Madison, Wisconsin area met several times over a three week period and wrote recommendations regarding nanotechnology research development — see their 'Report of the Madison Area Citizen Consensus Conference on Nanotechnology,' which was also submitted to you on Jan. 31st 2007.

Some of the Consensus Conference participants, in cooperation with faculty in the UW—Madison Nanoscale Science and Engineering Center created Madison's Nano Cafés,[1] which are public gatherings designed to educate participants and to provide an opportunity for engagement and critical thought about nanotechnology issues. The

---

[1]The Nano Cafés provide a casual atmosphere where non-scientists can hear from experts, ask questions, and share thoughts and concerns. UW-Madison experts explain their work, answer questions and hear ideas from members of the public. For details, see the dedicated website: www.nanocafes.org

citizens who have been involved in planning and organizing the Nano Cafés have formed their own group, the Citizens Coalition on Nanotechnology (CCoN).

CCoN members help organize Nano Cafés and make resources on nanotechnology available on our website: http://www.nanocafes. org. We believe that sharing different perspectives is essential to healthy public deliberation and democracy and want to have our say in nanotechnology research directions since these decisions and their outcomes will affect us.

We are grateful to be able to submit our comments and are deeply convinced that more funding is urgently needed for environmental, health, and safety research, which we believe should be prioritized as follows.

Environmental Health and Safety Priorities

Food Safety

We think safety of food should be the highest priority, particularly products which are on the market right now.

People are ingesting food with engineered nanomaterials in it right now. Engineered nanomaterials are in direct contact with the human body (including blood, organs, and other tissues). Scientists haven't studied whether or not these nanomaterials are safe or what their health effects might be over the long run. There are still many unknowns; we know very little about the materials used in these foods.

Because everyone eats food, food safety affects everyone, particularly children. If food containing nanomaterials turns to be harmful, it could have adverse effects on the perception of nanotechnology, setting off an overreaction that could stop or stall other nanotechnologies that could benefit humans.

We request that more research be done on:

- the nanomaterials that are already in food products on the market
- the potential of nanomaterials used in foods to cross the blood–brain barriers
- what kind of health effects nanomaterials in food products have on the digestive system

In the meantime, while this research is being done, we request that food products that include nanomaterials (either natural or engineered) be labeled.

Finally, nano food products that are currently in development should not be put on the market until more health and safety research has been done.

Safety of non-food consumer products that are already on the market — with a priority on those which involve ingestion or direct contact with the body (e.g., cosmetics, nanoceuticals, textile)

Many non-food consumer products containing nanomaterials are also currently on the market. Some of these are being inhaled, ingested or applied on the skin. Again, very little is known about the health effects of these nanoproducts.

Therefore, we think there should be a research priority on products that involve ingestion or direct contact with the body (e.g. cosmetics, nanoceuticals, and textile). More specifically, we think the following questions should be addressed as soon as possible:

- Do nanomaterials in products put on the skin get through skin?
- What long-term effects might these materials have?
- What kinds of health effects do products that are ingested (nutriceuticals) have on the digestive system?

In the meantime, while this research is being done, we request that consumer products that include nanomaterials (either natural or engineered) be labeled.

Finally consumer products containing nanomaterials that are in development right now should not be put on the market until more health and safety research has been done.

Environmental Releases

Nanomaterials in products that are on the market right now have already entered the environment and, if not, will in the future. Some nanomaterials are being used intentionally for environmental remediation. If they are in the air, the soil, and the water, they are likely to enter human bodies, affect wildlife, and they may also have irreversible eco-system effects.

Given this, we think that the following research questions should be high priorities:

- What environmental releases are occurring from nanoproducts already on market (e.g. nanosilver washing machines, textiles, cosmetics, nanoceuticals, sprays, aerosols)?
- What are their potential health and environmental effects?
- Are they biodegradable?
- Will they build up in food chain over time?

In order to do answer these questions, researchers need to develop safe sensors for environmental monitoring (air, water, soil) of nanomaterials as soon as possible. Developing these sensors should be a research priority. Releases and levels of nanomaterials in the environment should be tracked and monitored using existing methods and using improved ones once sensors are developed.

In the meantime, we request that methods be developed to prevent the releases of nanomaterials into the environment until more is known about their short and long-term health and environmental effects.

Occupational Safety

We think research on occupational safety should be a priority because people working with nanomaterials are exposed right now on a regular basis. We have no data on human exposures in labs and industries that produce and handle nanomaterials.

Exposure data is very important, since it would enable risk assessors to track exposures and potential health effects through time. It would also help in the development of health and safety protocols as nanomaterial production increases.

We think that the following research areas should be high priorities:

- Developing appropriate and inexpensive monitors for workplace monitoring of nanomaterials.
- More research and monitoring of human exposures and health effects of workers, particularly:
  - o in industries that currently produce nanomaterials (particularly nanopowders and nanofibers)
  - o in industries that have been making nanomaterials since the 1990s (e.g., fullerenes and nanotubes production facilities)

- Moreover, while this data is being gathered, health and safety protocols should be immediately developed in research labs and industries that handle nanomaterials.

Consumer Protection Bodies

People trust a democratic government to take care of citizens' safety. If our government does not do that, who will?

As US citizens, we are concerned that negligence of government for assuring public safety regarding nanotechnologies could permit disastrous results for us all, short term as well as long term.

All products containing nanomaterials should be tested for human and environmental safety by some appropriate agency (existing or newly created) before they are released onto market or into the environment.

This means that there should be adequate funding appropriated to accomplish that purpose.

Communication/Citizen Participation Research Priorities

People need to know more about these technologies in order to be able to make informed choices about them and to affect research policy in wise ways. That's why we think communication and citizen participation should be high priorities for more funding.

Information Availability/Communication

Currently, the general public largely lacks awareness and understanding of nanotechnologies. However, people have a right to be informed about new developments in science and technologies that affect them. In particular, people have a right to know what is in products they use and consume, and what their health and environmental effects might be, even before certainty about their potential risks is established. They should be given the opportunity to learn about and evaluate information on nanotechnology and make judgments accordingly, even if — all the more when — there are risk uncertainties that experts can't resolve right now.

More upstream communication about health and safety should be a priority, so that citizens' perspectives heard at a stage when they can still influence research priorities.

We request that information free of jargon and of acronyms be shared with the public and policymakers, in a way that is accessible to and understandable by non-scientists. In particular, we think that these outreach efforts must be geared towards

- the people who are directly affected by and in contact with nanomaterials: consumers and people already working with nanomaterials in labs and in industries.

In order to communicate with affected individuals, we need to know who and where these affected individuals are. Thus, we need to fund research to track where nanomaterials are being produced, how many workers are in contact with nanomaterials, and who is using these products.

More generally, the public needs more information on nanotechnology research and product development, and especially products that are on the market right now. So, in particular, we think the public should receive more information about the following issues:

- Where nanomaterials are being produced?
- How many workers are in contact with nanomaterials?
- Who is using these products?
- Which products contain nanomaterials, of which elements, at what dose, and what are the risks and the uncertainties associated with them?
- What is the purpose for research grants applied in publicly-funded research institutions?
- What are the potential risks of any products likely to result from nanotechnology research/applications?

In addition, we think that people responsible for making decisions and taking actions to address potential environmental and public health risks related to nanotechnologies should also receive information about the issues listed above and in the rest of this document. In particular, the following public officials should receive this information:

- legislators (at the state and the federal levels) who are and will continue making decisions about public health, environment, science policy;

- staff in federal *and* state/local government agencies responsible for protecting the environment and public health.

We also request that information be made available not only on governmental websites, but also in mass-media and conferences, newspapers, television, radio, magazines, websites, blogs, science museums, and any popular media.

We request the extension of 'whistle-blower' protection status for scientists and other experts who raise issues regarding nanomaterials hazards or potential hazards.

Public Engagement

People have a right to be involved in what their world and the world of their children will become. Science and technology have — and will continue to have — huge impacts on these worlds, but the public is rarely asked to participate in their making.

Government officials and researchers could learn a lot by talking with citizens, because they have pertinent knowledge and perspectives different from scientists, industries and government experts' perspectives. Citizen perspectives are based on their concerns for peace, human health, and environmental well-being, now and in the future. They are not representative of any interest groups and do not seek short-term profits, but the collective interest. This makes their additional perspectives very pertinent as priorities are set, leading to decisions that are more in tune with public values and interests. Further, the more diverse perspectives that are included (of citizens, scientists, government officials, and industries), the better decisions we can make.

By engaging with citizens, government bodies and researchers can find out if environmental, health, and safety research is relevant to people who are using, working with, and eating products with nanomaterials.

Moreover, by engaging and communicating with people, researchers and risk communicators can find out if the people who are most likely to be exposed to nanomaterials have access to information about potential risks related to these materials, whether or not they are getting the information, and what kind of risk information would be most useful and accessible for them.

Government should initiate and fund opportunities for citizen involvement that would include participation with experts in the nanotechnology fields. Scientists and other experts should be required to communicate with the public regarding their work on nanotechnologies through public forums such as town meetings, consensus conferences, workshops, Nano Cafés, and so forth.

Government should commit to and take into account the input and conclusions of citizen recommendations and reports that are developed in citizen engagement efforts. Government should incorporate more regular opportunities for ongoing citizen input into decision-making processes related to nanotechnology development as well as development of other technologies. More regulatory meetings should be open to the public and held in locations that are accessible to diverse participants.

Research Funding Priorities

We have outlined above the specific research and communication/ engagement areas we think should be priorities, but we want to emphasize, in conclusion, that given the limited funding allotted for nanotechnology health and safety research, the following areas should be priorities for research funding:

- While the potential for improving life on earth with nanotechnologies is high, the risks associated with some of the uses and applications should be assessed before allowing these materials to be marketed.
- Risk assessment on products already available to the public should be the highest priority for research funding.
- Risk assessment for individuals who are researching and working where nanomaterials are manufactured or used in industry should also be a very high priority. Funding should also help determine the precautions necessary to prevent potential harm to these individuals.

Finally, of the money that is spent on nanotechnology development, we request that there be a focus on developing technologies that improve human well-being and decrease the desire for war. Rather than spending excessively on the means to make war, develop means closer to society concerns so that people's desires can be accomplished cooperatively.

Thank you very much for providing the opportunity to comment on this important issue. We hope you will consider our comments seriously.

Sincerely,

The Citizens' Coalition on Nanotechnology
Madison, Wisconsin

Chapter 9

The U.S. National Citizens' Technology Forum on Human Enhancement: An Experiment in Deliberation Across a Nation

Fern Wickson,[a] Michael D. Cobb,[b] and Patrick Hamlett[c]

[a]GenØk - Centre for Biosafety, Forskningsparken in Breivika, P.O. Box 6418, 9294 Tromsø, Norway
[b]Department of Political Science, North Carolina State University, 223 Caldwell, Campus Box 8102, Raleigh, NC 27695, United States
[c]Science, Technology & Society Program and the Department of Political Science, North Carolina State University, United States
fern.wickson@genok.no

9.1 Introduction

A growing literature asserts the moral and pragmatic virtues of including ordinary citizens in decision-making about science and technology policies (Rogers-Hayden & Pidgeon, 2008; Willis & Wilsdon, 2004). This can stem from a normative rationale in which the involvement of lay citizens is simply seen as the most

Consumers and Nanotechnology: Deliberative Processes and Methodologies
Edited by Pål Strandbakken, Gerd Scholl, and Eivind Stø
Copyright © 2021 Jenny Stanford Publishing Pte. Ltd.
ISBN 978-981-4877-61-9 (Hardcover), 978-1-003-15985-8 (eBook)
www.jennystanford.com

appropriate approach to science and technology in democratic societies, or from a substantive rationale in which their involvement is thought to lead to substantively better decisions (Stirling, 2008). With the rise of nanotechnology, it has often been claimed that a unique opportunity exists for citizens to be engaged at an early stage in the innovation process, namely, before products have been commercialised and research trajectories set (Delgado *et al.*, 2010). This call for 'upstream' public engagement in nanotechnology development is often portrayed as an attempt to avoid the type of backlash that was directed against genetically modified foods, where public engagement took an 'uninvited' form (Wynne, 2007) and arguably came too late to influence the establishment of transgenics as a research trajectory for solving agricultural dilemmas. It can also be argued that 'upstream' public engagement on nanotechnologies allows for discussions to move away from a narrow focus on the risks associated with particular products and instead open up for broader debate on the visions and socio-technical imaginaries driving particular fields of research forward. In line with this call for early or upstream engagement in nanotechnology, the 2008 National Citizens' Technology Forum (NCTF) was designed to promote citizen understanding and participation in science policy regarding human enhancement before such technologies became commercialised in a consumer market (Hamlett *et al.*, 2008).

The NCTF was a month-long process in which groups of ordinary citizens deliberated simultaneously in six different locations across the United States on the topic of converging technologies for human enhancement. The locations for the deliberations were New Hampshire, Georgia, Wisconsin, Colorado, Arizona, and California. Researchers from a local university at each location served as coordinators and facilitators for the individual groups, while the process as a whole was initiated by the Centre for Nanotechnology in Society at Arizona State University (CNS-ASU) and co-ordinated by collaborating partners at North Carolina State University. At each of the six locations, panels of (maximum) 15 lay citizens were recruited to deliberate and develop recommendations about the future of converging technologies for human enhancement.

Before their first meeting, citizens were provided with selected background materials on the topic. The background document (NCTF, 2008) included various imagined future applications of converging technologies, including a list of 'twenty NBIC developments thought possible or likely in the near future' such as 'direct broadband interfaces between the human brain and machines...[w]earable sensors and computers enhancing individuals' awareness of health condition, environment, chemical pollutants, potential hazard... [a] human body more durable, healthy, energetic, easier to repair, and resistant to many kinds of stress, biological threats, and ageing processes,' etc. Prior to the start of the process, participants were also asked to complete questionnaires about their personal backgrounds, their knowledge and opinions on these technologies, and their levels of trust and efficacy. Following the completion of the NCTF process, citizens also completed a final survey measuring these same opinions and attitudes (citizens who applied to participate but were not selected served as a control group by also completing surveys before and after the NCTF).

Procedurally, the meetings were face-to-face interactions within the individual groups on the first and last weekend of the month, while interactions across the different groups occurred in 9 online sessions held throughout the month. In some of these online sessions, the participants (panellists) had the opportunity to pose questions they had formulated to experts in the field. At the end of the forum, citizens at each site generated an independent report with recommendations for policymakers on how to manage these new technologies. Importantly, the groups were asked to reach consensus about their recommendations rather than to vote on each point, although each group had some freedom as to how to manage minority opinions in the development of their consensus report.

This deliberative process was specifically focused on the range of technologies that are converging around a promise of enhancing human abilities (this means enhancement of both mental and physical abilities as well as an extension of lifespan). In the United States, these so-called converging technologies are seen as stemming from four different areas: nanotechnology, biotechnology, information technologies, and cognitive science and are often referred to as NBIC technologies.[1] As these technologies have not yet successfully converged to result in actual applications to enhance

---

[1]Interestingly, in a European Commission report, it has been contrastingly emphasised that the social sciences will also converge with these fields of science and that the convergence need not occur around the goal of human enhancement but can also be actively directed towards socially agreed priorities and needs (see Kjølberg *et al.*, 2008; Nordmann, 2004).

human performance and abilities, the deliberative process was not concerned with consumer goods but with developing informed and deliberative public opinions on an emerging field specifically before it has reached a stage of delivering commercial applications.

9.2 Description of the Process

The NCTF was designed for several purposes. One objective was to study the feasibility of citizen deliberation about science and technology issues like human enhancement. Some scholars question the wisdom of having ordinary citizens deliberate about social and scientific issues (Mendelberg, 2002), so the NCTF was intended as a test of whether 'average, non-expert citizens can understand even quite complex issues and, if they have adequate information, they can come to sensible, informed judgments about those issues (Hamlett *et al.*, 2008).' A second and related goal was to determine whether citizen deliberation could be structured to take place at a national scale in a country as geographically large as the United States. Thus, a key ingredient to the deliberations was the use of the Internet to permit citizens from all six locations to meet simultaneously many times during the process.

A third objective of the NCTF was to inform policy, with the idea being to connect the judgments of informed citizens to policy-making processes. One way the policy could be affected was through the dissemination of the consensus recommendations in the citizen reports to other citizens. As the organizers wrote, the NCTF could 'provide information to other concerned citizens about techniques like this one . . . to enhance the abilities of ordinary citizens to help shape public policy on important issues'. In addition, the NCTF was supposed to generate recommendations to be delivered to policymakers in a particular field of technology development. Describing the ideal engagement process as 'upstream' in the technologies' development, the organizers said 'the goal of this project was to present the *informed, deliberative opinions* of ordinary, non-expert people for the consideration of policymakers who are responsible for managing these technologies *before those technologies are deployed*' (italics in original; Hamlett *et al.*, 2008). In the NCTF, each of the six different working groups developed consensus-based recommendations for managers and policymakers on technologies to enhance human abilities.

While it was indicated during recruitment procedures for the NCTF that the process was specifically aimed at generating recommendations for policymakers on how to manage the emergence of converging technologies for human enhancement at an early stage in the technologies' development, there was in fact no formal or direct linkage between NCTF outcomes and relevant policymakers' decisions. It is therefore questionable how and whether the citizen recommendations can actually affect policy. Despite the NCTF's inability to guarantee a direct link to policymakers, it could be argued that policy might still be affected by the process through several different follow-on activities, including the indirect dissemination of citizens' reports. For one, the U.S. Congress had previously identified consensus conferences as a specific type of citizen engagement that should be encouraged in the field of nanotechnology when it authorised the US National Nanotechnology Initiative (P.L. 108-93). Building on the spirit of this legislation, Dave Guston at CNS-ASU coordinated a briefing of the nanotechnology caucus in Congress on March 9, 2009. At this briefing, researchers reported their judgments about the utility of the NCTF, as well as distributed summaries of the citizens' recommendations for action. Finally, participants at one of the sites (Colorado) took the initiative to write an open letter to their U.S. congressional representatives well after the NCTF had concluded, encouraging them to listen to their policy recommendations.

Despite these secondary avenues for influencing policy that have been pursued since the completion of the NCTF, it is worth noting that the way the process lacked a formal policy connection affected the feelings of the lay participants. While Cobb (2011) reports that those lay citizens involved in the process experienced increased feelings of internal political efficacy as a result of taking part in the process (i.e., I am capable of deliberating), panellists were also less likely to endorse expressions of external efficacy (i.e., policymakers listen to people like me). A likely explanation for these divergent feelings is that citizens learned by experience that their own capabilities were somewhat greater than they presumed, but they also believed nobody was going to listen to what they had to say. As a result, the NCTF became more of an academic exercise to participants, and this was not completely in line with the original objectives of the organisers (Delborne *et al.*, 2011; Kleinman *et al.*, 2011).

The process was funded as part of a grant awarded to CNS-ASU (and partners) from the National Science Foundation of the United States. In 2003, the U.S. Congress passed legislation authorising the National Nanotechnology Initiative (P.L. 108-93), and in the language of the act, upstream public engagement with technology was encouraged. More importantly, consensus conferences were specifically identified as a preferred method for involving citizens with science policy. In terms of implementing the NCTF as a particular exercise in citizen engagement, the initiator of the process was the CNS-ASU. The coordination and oversight of the project was, however, conducted by collaborators from North Carolina State University, Drs Patrick Hamlett and Michael Cobb. They had previously conducted a number of consensus conferences on other scientific issues (though none at a national level), and their expertise in organising and facilitating such events saw them take a primary role as partners in the initiative (Cobb, 2004; Hamlett, 2002; Hamlett & Cobb, 2006). The primary coordinator of the project was Dr Patrick Hamlett from North Carolina State University, an associate professor in the Science, Technology and Society Program and the Department of Political Science. His co-principal investigator, Dr Michael Cobb, is an associate professor in the Department of Political Science at North Carolina State University. Each of the six sites for the different participant groups had local facilitation teams coordinating their activities (see Appendix A for details).

Activities supported by the grant spanned three years, including the planning, implementation, and result analysis. Although the primary investigators (Hamlett & Cobb) were awarded approximately $240,000 to conduct the NCTF and analyse it, the total amount of money invested in the NCTF was over $500,000 after accounting for CNS-ASU support for additional researchers at each of the six locations.

Planning for the NCTF began in earnest one year before the process was conducted in March 2008. The majority of time in this period was spent developing the background document for participants to learn about the topic and the various survey instruments to measure the impact of deliberation on the participants' opinions. Researchers at each site location were encouraged to develop their own research agendas and to submit questions to be placed on the surveys. Additional time was used to plot citizen-recruitment strategies, prepare the researchers for the job of being a site facilitator for deliberations, and identify experts that could be invited to speak to the participants during the deliberation process.

The NCTF took place over the entire month of March in 2008. It began with members of the different working groups meeting face-to-face on the first weekend of the month (the 1st and 2nd of March) and concluded with another face-to-face meeting over the last weekend of the month (29th and 30th of March). In between these face-to-face meetings for each individual working group, the process involved nine online sessions (or keyboard-to-keyboard meetings). These were held on Tuesday, Thursday, and Sunday evenings throughout the month and ran for two hours at a time. During these keyboard-to-keyboard sessions, participants from the six different working groups were given the opportunity to hear the perspectives of the other groups and to interact with the participants from other locations across the country. Like the face-to-face meetings, these keyboard-to-keyboard meetings were also organised and supported by facilitators. The online sessions were particularly focused on sharing lists of important issues and questioning the invited subject matter experts. Interestingly, during these online meetings, participants were arranged into six different teams, with each team comprising a mix of participants from the various locations. At each online meeting, one team at a time was 'actively' taking part in the discussion while the other teams listened to/read the interactions. All teams were given a chance to actively contribute to the discussion, just not all at the same time. This rotating approach to active online communication was primarily put in place as an attempt to manage the online interaction in an effective way.

After the direct interactions of the NCTF ended, researchers have been analysing the survey data, writing papers, and presenting their results at conferences in the United States and in Europe. Most recently, Cobb and Gano (2012) collected longitudinal data about the NCTF participants' opinions to explore whether their involvement had had long-term consequences.

The participants in the process were primarily non-expert lay citizens from across the United States. Each of the individual sites recruited participants through newspaper and Internet advertising. Applicants were told that they would receive a $500 stipend for successfully completing the month-long process. Subsequent analysis of citizens' reasons for taking part in the NCTF indicates that financial compensation was an important, though not determinant, factor in their decision-making (Kleinman *et al.*, 2011). As part of the selection procedure, people wishing to participate in the process had to complete an online questionnaire with general questions about themselves (demographic information such as age, gender, education, ethnicity, political persuasion, etc.) as well as their responses to a set of knowledge and attitude questions relating to nanotechnology and human enhancement (how much have you heard about nanotechnology, what do you think about the risks and benefits, etc.). Answers to questions such as those about risks and benefits had no bearing on whether applicants were selected to be a participant or not and were instead used to provide baseline information about all the applicants so that the researchers could compare opinions held by those who did and did not go through the deliberative process. One of the roles of this initial questionnaire was to uncover any potential conflicts of interest (with questions like, are you financially invested in nanotechnology or have you participated in an NGO that has taken an active position on nanotechnology, etc.). Answers to these questions were then used to exclude some potential participants on the basis of a conflict of interests. Participants were chosen from the pool of applicants by facilitators at each site to specifically try and ensure that the participants were 'diverse and roughly representative' in a demographic sense (Hamlett *et al.*, 2008). Efforts were made to have participant groups match both local and, in aggregate, national demographics. While this was largely successful, both the applicants and final participants were more liberal and well educated than the U.S. population as a whole (Table 9.1).

In addition to the lay participants, experts in particular subject areas were also invited to participate in the process. Their involvement was to specifically answer questions that the participants formulated during their deliberations. These invited experts included Dr Roberta M. Berry (a specialist on the legal, ethical, and policy implications of life sciences research and biotechnology from the Georgia Institute of Technology), Dr Steven Helms Tillery [a specialist on cortical neuroprosthetics from Arizona State University (ASU)], Dr Maxwell J. Mehlman (a specialist in the federal regulation of medical technology from Case Western Reserve University), Dr Kristin Kulinowski (executive director of the Center for Biological and Environmental Nanotechnology at Rice University), and Dr Jason Scott Robert (a bioethicist and philosopher

Table 9.1 Demographic information on the lay participants involved in the NCTF

	Applicant	Panellists	National
Sex	42% Male	50% Male	49% Male
	58% Female	50% Female	51% Female
Education	25% 'Some college'	29% 'Some college'	50% 'Some college or a college degree'
	33% 'College degree'	31% 'College degree'	
	33% Grad school	31% Grad school	9% Grad school
Party ID	48% Democrat	44% Democrat	36% Democrat
	11% Republican	9% Republican	27% Republican
	30% Independent	36% Independent	37% Independent
Political Ideology	48% Liberal	41% Liberal	25% Liberal
	14% Conservative	14% Conservative	36% Conservative
	28% Moderate	27% Moderate	35% Moderate
Race	71% White	65% White	66% White
	16% Black	15% Black	12% Black
	5% Asian	6% Asian	4% Asian
	5% Hispanic	7% Hispanic	15% Hispanic
	<1% Native American	2% Native American	
Median Household Income	9% <$15K	9% <$15K	
	16% $15K–35K	21% $15K–34K	
	21% $35K–50K	16% $35K–50K	
	23% $50K–75K	20% $50K–75K	
	15% $75K–100K	16% $75K–100K	
	16% >$100K	17% >$100K	
Median Age	37 years old	39 years old	37 years old

Source: Reproduced with permission from Hamlett *et al.* (2008).

of science from ASU). The extent to which these experts covered the relevant range of opinions and specialist subject areas is debatable; however, the organisers reported having had a longer list of potential experts available for inclusion in the process but participants did not ask to hear them. It should also be noted that the above-listed facilitators and coordinators from universities at each of the sites also participated in the process. Conversely, no government officials from any level were involved. As this was a process directed towards lay citizens, the inclusion of all relevant stakeholders was not the primary aim of the process. Instead, the principal motivation was to present balanced information and to avoid having the process 'hijacked' by self-interested stakeholders.

9.3 Review of the Process

9.3.1 Initiatives and Objectives

In terms of creating informed public opinion about human-enhancement technologies, the NCTF succeeded. Compared with factual knowledge held by participants before deliberating, knowledge improved dramatically at the end of the process (Cobb & Hamlett, 2008). In addition, qualitative analysis of the consensus reports suggests a level of thoughtfulness that could not be matched by ordinary citizens who had not taken part in the NCTF (Hamlett *et al.*, 2008). Furthermore, opinion shifts that occurred about the desirability of human-enhancement technologies and their applications during the NCTF were unrelated to pathologies of small-group decision-making called polarisation cascades (Sunstein, 2000). This suggests participants changed their minds due to informational and deliberative processes and not due to social pressures or other undesirable reasons (Hamlett & Cobb, 2006).

On the other hand, it was important for the NCTF to create a linkage between citizens' policy recommendations and actual policymaking, and in this respect, the NCTF was not as successful. There is minimal evidence to suggest the citizens' reports have reached decision makers, and no examples to date that indicate any of these recommendations have been adopted. While almost all participants in the follow-up longitudinal study endorsed the idea of holding additional consensus conferences (Cobb & Gano, 2012), they also advocated linking them to policy making. One interesting finding from the longitudinal study is that participants of the NCTF were later on not more likely to be involved in politics or general issues about science and technology than non-participants who applied to the NCTF in 2008, but they were significantly more active regarding human-enhancement technologies in particular, such as talking to friends and family about the topic and visiting museum exhibits (Cobb & Gano, 2012). This supports the argument made by Powell and Kleinman (2008) that participation by ordinary citizens in consensus forums can be an empowering experience.

9.3.2 Organisation

The direct costs (salaries for faculty and students organising the events, payments for food for individual participants, advertising, materials, etc.) and indirect costs (mostly university-mandated overheads for facilities and administration) totalled approximately US$500,000–600,000. One of the members of the initiating group at CNS-ASU, Professor David Guston, has suggested that on one hand, this total under-estimates the amount involved. This is because some personnel and administrative costs were not separately budgeted for this activity, which relied on pre-existing infrastructure in some places. However, on the other hand, it can also be seen to over-estimate what was needed. This is because the funding body of the National Science Foundation requires grant holders to charge indirect costs on the first US$25,000 of subcontracts at the subcontracting institution, meaning that ASU counted its own (49.5%) overhead on the first $25K that went out to North Carolina State University (which oversaw the project) and each of the five other (non-ASU) partners, which all charged their own overhead on the whole of the subcontracts. This means that a different administrative set-up under different funding schemes might have been able to spend less on facilities and administration.

While the figure of US$500,000–600,000 certainly seems substantial, it should be remembered that this was the first attempt in the United States to create a national consensus conference style process across various locations in a simultaneous manner. Therefore, the planning and execution of this process was necessarily going to require a substantial funding investment. Given the scale of the operation, the initiators have indicated that they see the process as having been equipped with a reasonable amount of resources. In claiming this, they emphasise that the citizens' reports with recommendations for policy should not be viewed as the only outputs attained for these costs. They suggest that the background document and the final summary report both represent outcomes that contain a significant amount of expertise. They also point to various academic analyses of the process that have been published and are in preparation that will inform both the theoretical literature on deliberation and the pragmatic literature about encouraging citizen engagement with science and technology. Additionally, the learning of the various participants should also be viewed as a valuable outcome, as should the information gleaned about the trial of a national scale deliberative process.

One factor affecting how well a consensus conference meets its objectives is the quality of the facilitation of group deliberation. This means that it could be argued that, ideally, trained facilitators rather than interested academics should have led the discussions of the NCTF. However, the use of such facilitators would have also added to the financial cost of the event. Therefore there is a need to consider the extent to which the involvement of trained facilitators is necessary to meet the process objectives and to weigh this against the funds available. It is also interesting to note that the Internet component of the NCTF was problematic according to many participants. Future events that utilise the Internet to compensate for spatial distances between participants should spend time and money to ensure that the Internet platform used for hosting deliberations is efficient and user friendly. Likewise, researchers need to ensure that the online component does not encourage 'social loafing' among participants whose behaviour cannot be seen by their peers or the facilitator. This is important because some participants openly admitted to walking away from online discussions to conduct chores or to watch television, for example.

9.3.3 Participation

The content and the scope of the process were arguably well defined for participants. The focus of discussions on converging technologies for human enhancement was clear and explicitly declared in all materials promoting the process and recruiting participants. The expectations of participants were also arguably clearly communicated. For example, potential participants were told that they would be asked to do the following:

(a) Read and review background information describing the technologies involved in human enhancement

(b) Raise and discuss any concerns or issues they think are relevant to those tech

(c) Consider the concerns or issues raised by other members of their panel and by members of the other panels

(d) Question experts who can add information for them to consider

(e) Attempt to draw up a set of recommendations that their panel would want to present to policymakers as they decide how to manage these technologies

(f) Write a collective report explaining their panel's thinking and the recommendations they agreed upon

After being selected to participate in the process, participants were given the background reading material about the technologies of focus that also worked to clarify the specific topic of discussion and ensure that it was clearly communicated to all involved. While the actual process itself, with its combination of face-to-face and keyboard-to-keyboard meetings across the range of sites, was arguably more complex than other deliberative processes, there was a concerted effort to communicate the procedures on the website in as clear and concise a way as possible. It should be noted, however, that while the desire to generate recommendations for policy and decision makers is clear in the material for participants, the participants were arguably not adequately informed up front about the challenges facing the transmission and translation of their recommendations in actual practice.

While understanding the existence and extent of conflict in participatory exercises is important, it is not easy to judge the extent to which this existed in the NCTF. This is because while there was no direct evidence or indicators of overt conflict between participants, not all of the possible data (such as video of face-to-face deliberations) has been analysed to understand this element, so it is possible that conflict was subtle and is yet to be analytically uncovered.

9.3.4   Results of the Process

The formal outcomes of the process were the consensus report documents containing the recommendations developed by each of the six participating groups. These are meant to resemble a 'common voice' of the non-experts involved and are specifically not supposed to reflect stakeholder interests. These recommendations were aimed at elected officials, government policymakers, business leaders, and others responsible for making decisions about the development of converging technologies. Although each of the groups developed their consensus recommendations independently, the reports tended to share a common vision of the most important issues requiring policymakers' attention. For example, there was unanimous support across all six sites regarding concern over the effectiveness of regulations for NBIC technologies and the need to provide public information, including more public deliberative activities and K-12 education about NBIC technologies (Hamlett *et al.*, 2008). Likewise, there was near-unanimous support (five of the six sites) regarding concern about the equitable distribution of new enhancement technologies, the greater importance of therapeutic over enhancement research, the need for careful monitoring of such technologies and the development of international safety standards for them, and an emphasis on development of such technologies to maximise their benefits with both public and private investment.

In generating the final group reports with recommendations for policy and future management of NBIC technologies, developing a consensus position (within the individual groups) was the explicit objective. This is, for example, demonstrated in the final report from the California group that they named 'consensus report' (NCTF California, 2008) and the Wisconsin group which refers to their recommendations as representing 'the consensus of the 14 members' (NCTF Wisconsin, 2008). Interestingly, however, the Wisconsin group report states that, 'For the purposes of this document, consensus indicates not unanimous support, but the wisdom of the group without major objection' (NCTF Wisconsin, 2008). In the report from the Arizona group, while the vast majority of the recommendations seem to stem from a consensus position and are presented as 'We think/recommend/conclude', etc., there are two occasions in the report where it states that 'Some of us are concerned...' (NCTF Arizona, 2008). This indicates occasions when group consensus was not achieved and a minority opinion is being presented. Reading the final reports gives a sense in which although consensus and shared opinion was sought, the meaning and importance of this was differentially negotiated in each group. It should also be noted that despite the interaction across the groups in online sessions, 'no effort to reach a single consensus involving all six sites' was pursued (Hamlett *et al.*, 2008). Despite this, however, participants' recommendations were fairly consistent across the six sites.

While consensus rules were used in the creation of the recommendation reports, consensus rules have been criticised as exacerbating cognitive and affective pathologies of small group deliberation (Sunstein, 2000). Here the argument is that individuals feel a social pressure to conform to the initial position held by the majority within a group, and this feeling intensifies when consensus is required for arriving at group level decisions. To evaluate these concerns, participants were asked at the end of the NCTF if they personally endorsed only some, a fair amount, most, or all of the groups' recommendations. The vast majority of participants (90%) either agreed or strongly agreed with the final recommendations of their group and therefore a high degree of tolerated consensus appears to have been achieved (Hamlett *et al.*, 2008). Likewise, just 16% of participants reported a personal objection to some of their groups 'consensus' points and another 3% personally objected to 'many' of the major points in the final report (Hamlett *et al.*, 2008). Additional analysis of the policy opinion shifts that occurred after the deliberation process found little evidence that participants were unduly influenced by pre-deliberation majority opinion (Cobb, 2011, Cobb & Hamlett, 2008; Hamlett & Cobb, 2006).

The documents listing recommendations from each of the sites have been dispersed and distributed differently by each of the groups involved, with the local organisers and facilitators having some autonomy over this dissemination process. There is no publicly available information that details what actions have occurred for each group's set of recommendations; however, a final report that summarises the process and its national findings (Hamlett *et al.*, 2008) is in circulation amongst the collaborators and this document is publicly available on both Dr Hamlett's and the CNS-ASU websites. Conversely, there are a growing number of publications exploring the value of the NCTF (Delborne *et al.*, 2009; Kleinman *et al.*,

2011; Philbrick & Barandiaran, 2009). Facilitators from some of the individual sites are writing academic papers about their own site experiences as well as cross-site analyses. The pre- and post-process questionnaires are also being analysed by the collaborators at North Carolina State University and researchers at ASU in terms of what they reveal about the process itself and the opinions of the participants on nanotechnology and human enhancement. Finally, Cobb and Gano (2012) are beginning to report the results of a longitudinal study of NCTF participants' attitudes about human enhancement and their political and social behaviours that may have been influenced by their taking part in the deliberations. These on-going academic analyses of the process and its findings represent a significant outcome of the project in addition to the consensus recommendation reports that the project generated.

Beyond the formal result of consensus recommendations reports and the academically desirable result of a range of analytical papers, a more informal outcome of the process was the enhanced awareness of the field and sensitivity to important issues that was generated among the participants. This learning can be observed through documentation in the final questionnaire where participants' knowledge was measured and compared with their pre-deliberation knowledge. Knowledge gains documented through these questionnaires were impressive, especially once their self-assessed confidence in their answers is taken into account.

9.4 Deliberation Criteria

The NCTF process was limited to discussing converging technologies for human enhancement. This thematic focus was justified by the organisers referring to these technologies as having enormous potential, raising pressing social and ethical questions and being close at hand. These features generate the sense that there is both a need and scope for managerial decisions and that these decisions would do well to be informed about public opinion on the issues of concern. In this sense, the focus is broader than nanotechnologies, although these are an important strand in the thematic focus of converging NBIC technologies. At a pragmatic level, there was a need to keep deliberations focused on NBIC technologies and their applications. The limited time allotted for the NCTF to finish meant the core objective of drafting consensus reports about the topic precluded a more free-wheeling discussion entirely determined by the participants.

9.4.1 Reasoned Process

Scientific and reason-based arguments were presented to participants in the background document and in discussions with the invited content experts. The background briefing document was written by various researchers at CNS-ASU and was subject to review by an oversight committee comprising of Ida-Elisabeth Andersen from the Danish Board of Technology and David Rejeski from the Woodrow Wilson International Center for Scholars. Specifically scientific arguments were brought into the process by the experts invited by the organisers to answer the lay citizens' questions.

The citizen conversations were also presumably rational, although there was no formal mechanism to ensure or enforce that this was the case. However, the consensus conference structure and process are designed to promote deliberation as opposed to mere discussion (as for example may often take place in a focus group setting). Deliberation, in contrast to other forms of conversation, is a particular type of talking. Talking becomes deliberation, says Habermas (1996), when argumentation is welcomed, and it is based on a free and equal exchange. Moreover, deliberation requires individuals to weigh carefully the consequences of various options for action and how others will view these acts (Burkhalter *et al.*, 2002). Although there have been no formal attempts to measure the presence or level of deliberation that occurred, transcripts and video exist that could be used to conduct such analyses.

The variety of opinions that were included in the process was linked to the selection of participants. As previously indicated, the selection process aimed to achieve diversity among the participants that was roughly representative of local demographics. In small groups, representativeness is impossible along all of the relevant dimensions, but care was taken to maximise the diversity on more crucial dimensions, such as gender. To be sure, demographic diversity is not a guarantee, only a proxy, for differences of opinions but it is important to note that diversity was specifically pursued in the choice of participants.

Room for mutual learning and a change in positions across these diverse participants was actively encouraged by the nature of the interactive dialogue — both face-to-face and keyboard-to-keyboard. The use of questionnaires at the beginning and end of process allowed the organisers to document the amount of learning among the participants as well as any significant changes in their positions and attitudes. Analyses performed on these questionnaires have found that learning and opinion change did occur, with participants showing substantial increases in their knowledge of nanotechnology and human enhancement. Interestingly, sometimes this was correlated with dramatically lowered percentages of support for the technologies and their potential applications (Hamlett *et al.*, 2008).

With almost all of the materials and findings regarding the NCTF being available in academic publications or online at CNS-ASU website, or by request from the organisers, there is a strong claim in favour of the process being a transparent one. It was certainly transparent in terms of the public availability of the background material provided for the process, the discussions that occurred during the process, and the recommendations that were developed through the process. Although some materials are not as easily accessible, such as the survey instruments used in the pre- and post-tests, they are all freely provided upon request and posted to websites when requested. The only remaining question that the organisers do not currently publicly address is what has happened to each of the site recommendations (i.e., to whom they have been disseminated, and how?).

The NCTF was therefore arguably a largely transparently conducted process involving a diverse range of participants who were exposed to reasoned argumentation on the topic of concern. It could also be suggested that the process design encouraged participants to deliberate in a rational manner and create a space for mutual learning and a change in positions to take place.

9.4.2 Equal Participants

All the lay participants had the same role in the process — to discuss the issues they saw as important, formulate questions for clarification from experts in the field, and generate recommendations for future management and policy decisions. In this sense, all the participants in the process were equal. While there were some experts that were invited into the process, they were simply there to answer the questions of the participants and in this sense, did not have any power over the participants; except perhaps in the sense that their position as experts may have granted them a perceived authority of knowledge that may have consequently influenced opinions. Previous research about small group deliberation worries that 'disadvantaged group members', such as women and non-whites in the American context, will find their voices further marginalised in these interactions. Either by being numerically outnumbered, or because they are more hesitant to speak up and defend their positions. The fear in this case is that deliberation only serves to strengthen the argumentative positions of the dominant groups. However, no direct evidence has been uncovered to suggest that this occurred in the NCTF.

9.4.3 Consensus-Driven Process

Compared with other kinds of citizen engagement events, the NCTF uniquely emphasises consensus rules. While participants were not asked to make every decision via consensus (e.g., when deciding the order of agenda items to discuss participants could make such routine decisions though plurality or majority voting rules), the NCTF did push them to reach consensus about the items included in their group's final report. As noted above though, some variation did exist on how the term consensus was interpreted, with at least one group choosing to interpret the meaning as indicating 'not unanimous support, but the wisdom of the group without major objection' (NCTF Wisconsin, 2008). The design of the NCTF was inspired by the model of the consensus conference and an attempt to experiment with how this model might be applied to a geographically large and demographically diverse nation such as the United States. Therefore, it can be argued that the goal of having citizens draft consensus reports is partly what defined this particular forum for deliberation.

While some scholars warn that consensus rules can make it nearly impossible to arrive at successful deliberative outcomes, forums such as the NCTF can work to empirically test this claim. According to the organisers, the high quality of the reports suggests that consensus rules did not impede the participants from producing a successful document. Survey data about their feelings and opinions also support this conclusion. Cobb (2012) reports that twice as many participants, for example, strongly agreed with a statement endorsing the use of consensus rather than majority voting to resolve group conflicts. Furthermore, nearly all of the participants thought that their views were fairly represented in the final reports. In a one-year follow-up survey of most of the original participants, Cobb and Gano (2012) find just six (10%) of them disagreed with the statement, 'overall, the final report reflected my preferences'. Similarly, only four participants (7%) in that longitudinal survey reported objected to major points in the document. While one might hope to find that every participant agreed with all of the major points in document, majority voting rules would have arguably produced a larger cohort of participants feeling marginalised.

9.5   Summarising Appraisal

The NCTF process aimed to be deliberative by design. It was not intended to be a top-down event in which organisers merely disseminated information to passive participants. Instead, the NCTF aimed to be a more bottom-up process whereby citizens had some freedom to determine the course of events. Of course it could be argued that the process structure, the topic chosen, the provision of background material, and the selection of experts all represent avenues through which the organisers maintained and exerted an influence over the shape and content of the discussions, the participants did have freedom in where they took their discussions, what questions they posed to the experts, how they engaged in the online sessions, and most importantly, the content of their final reports. With assistance from facilitators, the participants set the agenda of their face-to-face meetings, determined the information that was necessary to locate, and talked with one another about their ideas and preferences.

While the literature on citizen engagement with science often conflates any kind of talking with deliberation about the issues, the NCTF was specifically designed to encourage deliberation, properly defined. Deliberation involves a particularly sophisticated version of talking, listening, and reasoning, which facilitators actively encouraged participants to engage in. Although participants' conversation have not yet been systematically analysed to measure the presence or level of deliberation, when arguing, participants were reminded to give reasons for their opinions, to listen to others perspectives, and to try and incorporate the opinions expressed by others in their responses. Encouraging this approach was an explicit task of the facilitators. Furthermore, efforts were made at the recruitment stage to generate demographic diversity among participants as proxies to include diverse ideas, and in this way, the deliberation arguably had to take place across diverse positions.

Compared with other kinds of citizen engagement forums, the strengths of NCTF process were the extended time frame (one month), the multiple locations engaged simultaneously, the combination of face-to-face and keyboard-to-keyboard interactions, and the generation of final consensus reports with recommendations. Although less obvious, another strength of the NCTF was its focus on a particular field of technology development and the engagement with this field at an early stage of development so that, theoretically, citizen input had a greater chance of acting on policy. A third aspect to the NCTF that compares favourably with other engagement forums is its more rigorous measurement of the various kinds of process outcomes, such as survey questions asked before and after deliberation that were designed to assess the effects of the deliberative process on the attitudes and opinions of the participants.

The principal weakness of the NCTF, however, is that it lacked a clear avenue for the outcomes of the process to affect policy and/or research and development trajectories. This limitation is arguably the reason that after the event participants reported a slight decrease in their belief that their opinions or actions can actually affect political outcomes (Hamlett et al., 2008). Secondly, the in-depth deliberations necessarily limit the number of participants that can be successfully engaged at one site at any one time. Adding more site locations, while possible, would have cost more money and yet still fail to significantly increase the number of participants since each group was limited to a maximum of 15 participants. The substantial financial cost of running consensus forums like these is arguably a limitation that will prevent them from being regularly and more widely used as a tool for citizen engagement. Finally, the final reports could have been structured differently to make minority viewpoints more prominent, although doing so would have likely encouraged participants to defect from the consensus opinions, making the process of crafting final reports more volatile and cumbersome.

The NCTF experience offers a number of valuable lessons about citizen engagement processes. First, the NCTF demonstrated that it is possible to hold deliberative engagement forums that bridge large geographic divides by using a combination of face-to-face and keyboard-to-keyboard meetings. The addition of the Internet to the deliberative process points to ways that online interactions might be included in similar processes. While the use of an online interactive component is something that can be modelled and utilised on other occasions, it arguably needs improvement to be maximally effective. In the case of the NCTF, very few participants (3%) said they favoured the online approach to group communication over face-to-face interaction (Hamlett *et al.*, 2008) This may indicate that face-to-face interaction should be seen as a preferred means of conducting deliberative processes, and yet, it is not certain that this should always be the case. For example, a number of obstacles to effective online deliberation in the NCTF can be traced to the particular Internet technology that was used rather than something inherent to all online deliberation methods. Future experimentation and evaluation of the use of online tools is needed to better understand their potential for facilitating citizen deliberation on matters of science and technology.

A second valuable lesson from the NCTF experience is that ordinary citizens are capable of deliberating about complex technologies and producing coherent policy recommendations about them. On average, citizens who participated in the NCTF emerged more knowledgeable, trusting, and internally efficacious. Their final reports, based on consensus rules, were thoughtful and reasoned. This suggests scholars looking to promote citizen engagement should compare more and less deliberative processes to determine whether the outputs of less deliberative forums are competitive and scholars should also work to ensure that engagement forums have direct connections to policy (for a comparison of lay public opinion about these technologies, see Hays *et al.*, 2012).

Thirdly, the NCTF demonstrated that a deliberation can lead to opinion change among members of the population that is independent of feared pathologies of decision-making processes in small groups. Instead of deliberations simply succumbing to the influence of positions held by numerical majorities or to the desires of a few vocal participants from more-privileged backgrounds, opinion change in the NCTF seems to have occurred for normatively desirable reasons. Participants became more critical of human-enhancement technologies, and preferred better regulatory oversight of them, due to information and deliberation. The fact that participants became more concerned about these technologies, but not fearful of them, suggests that it is desirable to provide balanced information about the risks and benefits of new technologies. Rather than hiding the risks, it should be realised that risk communication will not inevitably result in panic. Policymakers should consider that by giving information and opportunities for deliberation across diverse positions, citizens can make reasonable recommendations and decisions on even the most complex of issues.

A final element of the process that other engagement forums should take notice of is the use of pre- and post-deliberation surveys. These allowed the conference organisers to specifically see and map processes of learning in the participants, any changes in attitudes, the presence or absence of various deliberation pathologies, and the levels of digression from claimed consensus positions. The rigorous measurement of deliberative outcomes not only provides interesting data for research and analysis about the desirability of citizen engagement, but it also represents a model for evaluating citizen engagement forums. Going forward, there is a growing need to explicitly evaluate engagement processes and to do this in a way that allows for outcome comparisons, for example, between the NCTF, consensus conferences in general, and other kinds of deliberative forums. The NCTF research design offers a model for making these kinds of evaluations. As public engagement and the facilitation of a range of deliberative exercises expands for nanotechnologies, conducting careful and comparative evaluations becomes a crucial step if we are to better understand the best use of scarce resources for these kinds of endeavours. The NCTF is just one of a vast range of experiments being conducted on how to engage the public in the advance of nanoscale sciences and technologies. It is therefore important that we are able to critically evaluate such processes, compare their strengths and weaknesses, and see their successes and failures. This process of critical and comparative evaluation is crucial if the process of democratising science is to advance in a fruitful and mutually beneficial manner.

References

Burkhalter S, Gastil J, Kelshaw T (2002) A conceptual definition and theoretical model of public deliberation in small face-to-face groups, *Commun Theor*, 12, 398–422.

Cobb MD. Citizen deliberation and democratic theory: a quasi-experimental examination of small group deliberations. Paper presented at: Annual Meeting of the Midwest Political Science Association; April 15–19, 2004, Chicago.

Cobb MD (2011) Creating informed public opinion: Citizen deliberation about nanotechnologies for human enhancements. *J Nano Research*, 13, 1533–1548.

Cobb MD (2012) Deliberative Fears: Citizen Deliberation about Science in a National Consensus Conference, in Kieran O'Doherty and Edna Einsiedel, eds), New York, Elsevier.

Cobb MD, Hamlett P (2008) The first National Citizens' Technology Forum on converging technologies and human enhancement: Adapting the Danish consensus conference in the USA. Paper presented at 10th Conference on Public Communication of Science and Technology, June 25–27, 2008; Malmö, Sweden.

Cobb, M.D., Gano, G. (2012). Evaluating Structured Deliberations about Emerging Technologies: Post-process participant evaluation, *Int J of Emerging Tech and Society*, 10, 96–110.

Delborne JA, Anderson AA, Kleinman DL, Colin M, Powell M (2011) Virtual deliberation?: prospects and challenges for integrating the Internet in consensus conferences, *Public Underst Sci*, 20, 367–384.

Delgado A, Kjølberg KL, Wickson F (2010) Public engagement coming of age: from theory to practice in STS encounters with nanotechnology, *Public Underst Sci*, 20(6), 826–845. Available at http://pus.sagepub.com/cgi/rapidpdf/0963662510363054v2

Habermas J (1996) *Between Facts and Norms: Contributions to a Discourse Theory of Law and Democracy* (W. Rehg, Trans.). Cambridge MA, MIT Press.

Hamlett P (2002) Adapting the Internet to citizen deliberations: Lessons learned. In: *Proceedings: Social Implications of Information and Communication Technology, IEEE International Symposium on Technology and Society*. Raleigh, NC: Institute of Electrical and Electronics Engineers, 213–218.

Hamlett P, Cobb M (2006) Potential solutions to public deliberation problems: structured deliberations and polarization cascades, *Policy Stud J*, 34(4), 629–648.

Hamlett P, Cobb M, Guston D (2008) *National Citizens' Technology Forum Report* CNS-ASU (Report # R08-0002).

Hays SC, Miller A, Cobb M (2012) 2008 National Nanotechnology Survey on Enhancement Results Overview, in *Yearbook of Nanotechnology in Society, Volume 2: Nanotechnology, the Brain, and the Future* (Sean Hays, Jason Robert, Clark Miller, Ira Bennett, eds), New York, Springer.

Kjølberg K, Delgado-Ramos GC, Wickson F, Strand R (2008) Models of governance for converging technologies. *Technol Anal Strateg Manage*, 20(1), 83–97.

Kleinman DL, Delborne JA, Anderson AA, (2011) Engaging citizens: The high cost of citizen participation in high technology, *Public Underst Science*, 20(2), 221–240.

Mendelberg T (2002) The deliberative citizen: Theory and evidence, in *Research in Micropolitics: Political Decision Making, Deliberation and Participation* (Delli Carpini MX, Huddy L, Shapiro R, eds), Greenwich CT, JAI Press.

National Citizens' Technology Forum (NCTF) (2008) *Human Enhancement, Identity and Biology: NCTF Background Materials*. Available at http://www4.ncsu.edu/~pwhmds/index.html#who

Natonal Citizens' Technolgy Forum (NCTF) Arizona (2008). *National Citizens' Technology Forum on Human Enhancement, Identity & Biology: Arizona Panelists' Final Report*. Available at http://www4.ncsu.edu/~pwhmds/Arizona%20Final%20Report.pdf

National Citizens' Technology Forum (NCTF) California (2008) *National Citizens' Technology Forum: Converging Technologies for Human Enhancement – Report of the California Delegation*. Available at http://www4.ncsu.edu/~pwhmds/California%20Final%20Report.pdf

National Citizens' Technology Forum (NCTF) Wisconsin (2008) *National Citizens' Technology Forum: Final Report from Madison, Wisconsin*. Available at http://www4.ncsu.edu/~pwhmds/California%20Final%20Report.pdf

Nordmann A (2004) *Converging Technologies — Shaping the Future of European Societies*, 2004, European Commission, Brussels. Available at ftp://ftp.cordis.europa.eu/pub/foresight/docs/ntw_report_nordmann_final_en.pdf

Philbrick M, Barandiaran J (2009) The National Citizens' Technology Forum: Lessons for the future, *Sci Publ Policy*, 36(5), 335–347.

Powell M, Kleinman DL (2008) Building citizen capacities for participation in nanotechnology decision-making: The democratic virtues of the consensus conference model, *Publ Underst Science*, 17(3), 329–348.

Rogers-Hayden T, Pidgeon N (2008) Developments in nanotechnology public engagement in the UK: Upstream towards sustainability? *J Clean Prod*, 16(8/9), 1010–1013.

Stirling A (2008) 'Opening Up' and 'Closing Down': Power, Participation and Pluralism in the Social Appraisal of Technology, *Sci Technol Hum Val*, 33(2), 262–294.

Sunstein C (2000) Deliberative trouble? Why groups go to extremes, *Yale Law J*, 110, 71–119.

Willis R, Wilsdon J (2004) *See Through Science*, London, Demos.

Wynne B (2007) Public participation in science and technology: Performing and obscuring a political-conceptual category mistake, *EASTS*, 1, 99–110.

Appendix 3

Facilitators

In New Hampshire the facilitation team included Professor Tom Kelly (Director of the University Office of Sustainability, University of New Hampshire) and Elisabeth Farrell (Program Co-ordinator, Culture & Sustainability, and Food & Society Initiatives, University of New Hampshire). In Georgia the facilitators were Professor Susan Cozzens (Associate Dean of Research, Ivan Allen College, Georgia Institute of Technology) and Ravtosh Bal (Graduate Assistant, School of Public Policy, Georgia Institute of Technology and Georgia State University). In Wisconsin those responsible were Professor Daniel Kleinman (Department of Rural Sociology, University of Wisconsin-Madison) and Dr Jason Delbourne (Post-doctoral Research Associate, Holz Center for Science and Technology Studies, University of Wisconsin-Madison). In Colorado the facilitation team consisted of Professor Carl Mitcham (Director Henneback Program in the Humanities, Colorado School of Mines) and Assistant Professor Jennifer Schneider (Liberal Arts and International Studies, Colorado School of Mines). In Arizona the team included Professor David Guston (Political Science and Director of the Center for Nanotechnology in Society, Arizona State University), Assistant Research Professor Cynthia Selin (Center for Nanotechnology in Society, Arizona State University) and Roxanne Wheelock (Graduate Assistant Centre for Nanotechnology in Society, Arizona State University). Finally, in California the facilitators included Assistant Professor David Winickoff [Bioethics and Society, University of California (Berkeley)], Mark Philbrick [Graduate Assistant, Department of Environment and Management, University of California (Berkeley)] and Javiera Barandiaran (Graduate Assistant, Goldman School of Public Policy).

PART III
STAKEHOLDER-ORIENTED DELIBERATIVE
PROCESSES

Introduction to Part III

Pål Strandbakken
SIFO Consumption Research Norway, Oslo Metropolitan University,
Postboks 4, St. Olavs plass 0139 Oslo, Norway
pals@oslomet.no

In this part we analyse a set of more heterogeneous deliberations and exercises in participatory democracy. Some are clearly "stakeholder oriented" (Chapters 11 and 12), while the first (Chapter 10) might just as well fit under a possible tag like "deliberative processes within research processes", which, by the way, could also include the Nanoplat experience (Chapter 12). The main point probably is that they all differ from the citizen-oriented, "representing the voice of the common man", deliberations analysed under Part II.

We still draw on Cohen's deliberation criteria, but the accounts of the events are not as clear and uniform as the ones presented in Part II.

The idea of bringing stakeholders into the decision-making process and into more or less soft regulation is an integral part of the thinking behind governance. In an emerging technology context, this is something quite different from representing the common voice. We move from situations where lack of prior knowledge and lack of interest (or "stake") in the phenomenon is an asset, to situations where actors with different, but higher, levels of expertise represent the interests of business, the research community (or certain parts of it), or the interests of NGOs.

Deliberations done as parts of research projects might appear to be very similar to the ones analysed in the previous part. Focus group studies of citizens' reactions to, and opinions on nanoscience and nanotechnology have been arranged all over Europe (and beyond), and they very much follow the same logic and procedures as the citizen-oriented deliberations. The main, perhaps even the only difference is formal: a real deliberation comes with a set of expectations of making a difference, politically. Either if it is arranged on behalf of political authorities in order to broaden the knowledge base of the decision making, or if it is arranged by NGOs wanting to introduce a democratically based set of opinions into the public debate, the rationale is political.

For research projects, the political impact is coincidental, based on the subsequent use of the material researchers. From a deliberation we at least expect some kind of minimal attempt at political impact; from a research project we expect insight and knowledge in the form of reports and articles.

For the chapter on standardisation (Chapter 11), the material we use, and our involvement with the phenomenon is rather old, but we feel that it remains relevant because it points to the *deliberation aspect of standardisation work*, and that is not necessarily time bound. In addition, the chapter was integrated in the overall design of the first edition of *Consumers and Nanotechnology*, even if it has aged.

The chapter on the online platform (Chapter 12) has also aged, mainly due to the very rapid development of online tools and digital platforms. It is obviously not state of the art technologically anymore, and if that was the only purpose, it might as well be deleted. It is interesting as a historical document, however, as the 'finalisation' of the Nanoplat project. In a way it was the platform that 'plat' in the Nanoplat title referred to.

In addition, the very active online moderation by SDS in the tests is worth looking at. In the end it was more a test of the situation, than a test of technologies, and as such we see it as remaining relevant.

Chapter 10

Experiments with Cross-National Deliberative Processes Within FP6 and FP7 of the European Union: The Convergence Seminars, the DEMOCS Card Games, and the Nanologue Project

Enikő Demény,[a] **Judit Sándor,**[a] **and Péter Kakuk**[b]

[a]*Center for Ethics and Law in Biomedicine, Central Europan University, Nador u. 9, 1051 Budapest, Hungary*
[b]*Department of Behavioral Sciences, University of Debrecen, 4032 Nagyerdei krt. 98, P. O. Box 45, Debrecen, Hungary*
demenye@ceu.hu

10.1 Introduction

Fostering the use of deliberative processes in the governance of science is one of the strategic goals of the European Commission's (EC's) science policy. This goal is formulated explicitly both in the EC's Action Plan on Science and Society (AP on SAS) (EC, 2001)

Consumers and Nanotechnology: Deliberative Processes and Methodologies
Edited by Pål Strandbakken, Gerd Scholl, and Eivind Stø
Copyright © 2021 Jenny Stanford Publishing Pte. Ltd.
ISBN 978-981-4877-61-9 (Hardcover), 978-1-003-15985-8 (eBook)
www.jennystanford.com

and in its Strategy and Action Plan on Nanotechnology (EC, 2004, 2005). According to AP on SAS adopted in 2001 if citizens and civil society are to become partners in the debate on science, technology, and innovation it is not enough to simply keep them informed, but they must also be given the opportunity to express their views in the appropriate bodies (EC, 2001 AP on SAS:14). According to the action plan, science activities shall focus on the needs and aspirations of Europe's citizens and initiatives that provide a space for informed debate on important issues of public concern by bringing together the public, interest groups, and policymakers. Such initiatives are thought to pave the way for sound policies thus complementing the formal decision-making process (EC, 2001, AP on SAS:14).

The EC's Strategy and Action Plan on Nanotechnology adopted in 2004 also includes among its strategic goals the fostering of deliberative processes. The EC's "integrated, safe, responsible" and "enabling" approaches to nanotechnologies (NTs) (EC, 2004) are seen as indispensable to the open, traceable, and verifiable development of NT, according to democratic principles (EC, 2004:18). The aim of the EC is to enable the safe development and use of nanotechnology and nanoscience (N&N) and ensure that the public can benefit from the innovations that they may bring while being protected from any adverse impacts. According to the EC's Strategy and Action Plan on Nanotechnology:

"Without a serious communication effort, nanotechnology innovations could face an unjust negative public reception. An effective two-way dialogue is indispensable, whereby the general publics' views are taken into account and may be seen to influence decisions concerning R&D policy. The public trust and acceptance of nanotechnology will be crucial for its long-term development and allow us to profit from its potential benefits." (EC, 2004:19).

Different approaches, both regulatory and non-regulatory, are being pursued to achieve these strategic goals, and the progress towards the achievement of these goals is reported every two years in the so-called implementation reports (EC, 2007; CEC, 2009).

In order to integrate the societal dimension of NT, the EC encourages dialogue with EU citizens/consumers to promote informed judgement on NT research and development (R&D) based on impartial information and the exchange of ideas (EC, 2005, 2009:11). The EC seeks to create the conditions for and pursue a true dialogue with all the stakeholders concerning N&N. To achieve these goals, the EC has several instruments at hand, one of them is the Code of Conduct on Responsible Nanosciences and Nanotechnologies, which explicitly states that it was meant to be used also "as an instrument to encourage dialogue at all governance levels among policymakers, researchers, industry, ethics committees, civil society organizations, and society at large with a view to increasing understanding and involvement by general public in the development of new technologies" (EC, 2009:12).

In this chapter we concentrate our attention on the EC Framework Programmes for Research and Technological Development, FP6 and FP7, and will overview and evaluate a number of selected deliberative processes on NT, funded under these schemes. These projects are not intended to substitute the member states' efforts in this field but aim "to facilitate cooperation between stakeholders in different Member States, to create a critical mass for topics of European concern, to identify national or cultural differences between regions and Member States or to play the role of a pathfinder for new developments" (EC, 2008:10).

While the FP4 and FP5 programmes have already funded a good number of NT projects, only in FP6 (2002–2006) and then in FP7 (2007–2013) has NT been identified as one of the major priorities (EC, 2004:8). With this change the number of projects focusing on ethical, legal, and social aspects (ELSA) and governance of NT has also increased from only one FP5-funded project that directly referred to governance in NT (Nanoforum, the Pan-European Forum for NT) to about 20 projects in FP6 (EU, 2008:10,11). The funding for NT-related ELSA projects have also increased from less than €500,000 in 2002 to €3,000,000 in 2008 (EC, 2008:13). According to the report on European activities on ELSA and governance of NT, most of these activities are FP scheme–specific support or coordination actions, primarily designed to explore and experiment with — at a European level — best practices in the fields of scientific advice, risk governance, and civil society participation. Some of these projects are directly focused on NT, others are treating NT as a case study, and some of the projects can be applied to NT too (EC, 2008:11).

This chapter will concentrate on a specific part of these projects, namely on those that aimed to foster the development of deliberative approaches in the field of NT. Our review does not include some of the important participatory events on NT that have been also funded under the FP6 scheme, since their character and scope differed from that of the processes organised in the framework of larger projects.[1]

In the framework of the Nanoplat project we have overviewed three deliberative processes [two of them were organised in the framework of Nanobio-RAISE (Nanobiotechnology: Responsible Action on Issues in Society and Ethics) research project, and one in the Nanologue project], and access to information was one of the main criteria of selection, since some of the projects concerned were still running at the time of evaluation or projects reports were not yet available.[2] In this chapter, while we have described the three processes overviewed during the Nanoplat project in more detail, we have done a comparative evaluation of other processes also.[3]

The primary materials we have used are the project reports and websites. Attempts to collect additional information about the processes, for example, via e-mails or phone, were unsuccessful. Since we did not have resources to conduct interviews with the organisers and participants of the reviewed processes, the evaluation we conducted is mainly based on the comparison of the project's stated objectives and its reported outcomes.

In addition, we had to take into account that different meanings granted to public engagement are strongly connected to competing definitions of NT and citizenship. Therefore, each vision of public

---

[1]Such participatory events were NanoTechYoung, November–December 2003; EuroNanoForum 2003 (nanotechnology in general), December 2003; EuroNanoForum 2005 (focus on nanomedicine), September 2005; Consultation on the Code of Conduct for Responsible Nanoscience and Nanotechnologies, July–September, 2007.

[2]**Nanobio-RAISE:** DEMOCS Card Game and the Convergence Seminars, http://files.nanobio-raise.org;
Nanologue: Europe-wide dialogue on the ethical, legal, and social impacts of nanotechnologies, www.nanologue.net. The reviews have been performed in the period September–November 2008.

[3]The other projects we will focus on are the following FP6 Projects: **Deepen** — Deepening Ethical Engagement and Participation with Emerging Nanotechnologies, http://www.geography.dur.ac.uk/projects/deepen/Home, more information about the project in Davis, Macnaghten and Kearnes (eds) (2010), Macnaghten *et al.* (2010) and Rip and Egan (2010); **NanoCap** — Nanotechnology Capacity Building NGOs and trade unions, http://www.nanocap.eu, more information about the project in Broekhuizen and Schwartz (2010); **Nanodialogue** — Enhancing dialogue on nanotechnologies and nanosciences in society at the European level; and the following FP7 projects: **Nanoplat** — http://nanoplat.org, see more about the project in Stø *et al.* (2010); **Framingnano** — http://www.framingnano.eu, see more about the project in Mantovani and Porcari (2010).

engagement is related to a particular model of understanding of how the citizen should engage in the governance of science and technology (Fischer, 2000; Kearnes & Wynne 2007). Funding, designing, or organizing a particular public engagement mechanism are ways to articulate visions of how the citizens should engage in science policy (Laurent, 2008:3). Since the initiatives we reviewed are parts of the EC's larger efforts on science policy, and on the promotion of deliberative processes in the field of science and technology, in addition to the evaluation conducted with reference to the projects' own objectives, the processes can be understood and interpreted against this larger background which "frames" the reviewed projects.

Two of the reviewed processes, the *Convergence Seminars* and the *DEMOCS (DEliberative Meeting Of CitizenS) Card Game* have been organised as parts of the Nanobio-RAISE research project. This project brought together nanobiotechnologists, ethicists, and communication specialists with the aim to anticipate the societal and ethical issues likely to arise as nanobiotechnologies (NBTs) develop, and secondly, to use the lessons from the European GM debate to respond proactively and responsibly to the probable public, media, and political concerns. The Convergence Seminars as a specific form of public-opinion focus group was one of the activities organised in the framework of the project and it intended to explore public views on NBT and to clarify the potential ethical issues emerging from their development. The 2½ hour sessions enabled 6–15 participants to discuss different paths of technological development. The seminars with civil participants were performed in four different countries of the European Union (EU). The other deliberative process organised in the framework of the project, the DEMOCS Card Game, is a policy-making tool that enables small groups of people to engage with complex public policy issues. The DEMOCS Card Game on NBT has been developed in the framework of the Nanobio-RAISE project and has been played by different groups in the UK and the Netherlands.

The third process we reviewed was the Nanologue project, which intended to prepare the ground for enhancing the dialogue on ethical, legal, and social aspects of NTs. The Nanologue project was a dialogue on NTs that brought together a wide range of technological experts and some civil society representatives. The dialogue was driven by the need to understand ethical, legal, and social implications of NTs and also to communicate this understanding by raising awareness and providing information to societal actors. The project included several participatory processes that resulted in the construction of three near-term future scenarios for NT development in Europe at 2015 and an ELSA of NT questionnaire called *Nanometer*. Some recommendations on future dialogues have been formulated as well. The three scenarios might become useful tools in planning deliberative processes, while the Nanometer could be used as a simple tool supporting industry and academia in reflecting on the ELSA of specific applications of NT.

In the following section, we will present these three deliberative processes in terms of their objectives, resources, organisers, participants, methods applied, and their results.

10.2 Description of the Process

10.2.1 The Convergence Seminars

The Convergence Seminars were organised by one of the Nanobio-RAISE project's partners, the Stockholm-based Philosophy Division at the Royal Institute of Technology. The seminars were funded from the budget of the Nanobio-RAISE project (€553,845), which is the 6th Framework Programme Science & Society Co-ordination Action funded by the EC. The Convergence Seminars were conducted in 2006 as one of the first public engagement processes that explicitly focused on NBT and its potential applications. The methodology and idea of Convergence Seminars were developed by the organiser of the process.

The Convergence Seminars took place at four locations as a one-day event in each location between 4th May and 6th December 2006.[4] The general purpose of conducting the Convergence Seminars was to explore public views on NBT and to clarify the potential ethical issues that might emerge from their development: identify some key issues for the public and specify their recommendations to policymakers. A convergence seminar is a new form of public participatory activity that involves structured decision-guiding discussions employing *hypothetical retrospection*. For that purpose, it is necessary to achieve

---

[4]The first seminar was hosted by Gotland University, Visby, Sweden; the second by the School of Law at the University of Sheffield, UK, on 28th July; the third by the Maria Curie-Sklodowska University, Lublin, Poland, on 25th November; and the fourth by the Institute for Molecular and Cell Biology in Porto, Portugal.

several criteria. A number of scenarios should be seriously discussed, each of which explores possible future consequences of decisions that we take now. A comparative and concluding discussion takes place in which each of these scenarios is taken into account. Furthermore, the methodology should be practical and the procedure should be easy to apply and performable in a few hours. A set of concrete scenarios should be developed for the discussions. These scenarios should be constructed in accordance with the requirements for hypothetical retrospection. Hence, they should bring us to some future point in time, but each scenario should lead us to a different "branch" of future development. Each scenario should describe in outline the respective branch in its full length and not just the "final" state at the point in time at which the hypothetical retrospection is enacted. The focus should be on some decision in the present or near-present time that the participants are asked to evaluate from the viewpoint of their scenario. Furthermore, different scenarios should be selected so that they represent, for different alternative decisions, branches in which these decisions give rise to problems that make them difficult to defend in hypothetical retrospection.

Ideally, a large number of scenarios should be included in the procedure. The standard procedure proposed and tried out for Nanobio-RAISE employs only three scenarios. For the same reason, participants are divided into groups and each scenario is discussed in detail by one group only. This first phase of one-scenario discussions has to be followed by procedures in which participants from the different groups in the first phase exchange experiences. This is initially done in small groups that are formed by regrouping the participants so that each new "convergence" group includes a representative from each of the scenario groups. In the third and final phase, the participants are assembled for a concluding discussion about what advice the participants would like to give to decisionmakers who will decide on the development of NBT.

The *scenario narratives* used in the workshop were based on material taken from scientific articles and reports on nano- and biotechnology and from fellow researchers' works in these areas and in ethics. Researchers experienced in working with scenario workshops were also consulted. The hypothetical decision-making situation was placed in the short-term future in year 2010; and the three scenarios, each representing a different course of development, were placed in year 2020.

The discussed three scenarios represented diverging lines of development in terms of precaution and progress and contained different ethical themes such as justice and distribution, privacy, health, and enhancement.

The participants of the Convergence Seminars were reached through various channels, providing the organisers with a group of 7–13 voluntary participants at each location. The same moderator facilitated the seminars, except one place, Porto. No monetary compensation was offered to the participants but coffee and a light meal was provided at the end of each seminar.

In the *first seminar*, eight participants in the group were members of local branches of the Swedish Society for Nature Conservation and "Fältbiologerna", a youth environmental organization. The seminar was held in Swedish and the scenarios and seminar discussions were translated to and from English. The *second seminar* had 12 participants, who were students or members of a science discussion club. The *third seminar* had a group of 13 participants, who were mainly students in linguistics, architecture, and chemistry. There were also two senior researchers from NT-related fields. The *fourth and final* had a group of seven participants, who were students or non-academic staff such as administrators. The entire seminar was held in Portuguese. The scenarios were translated in advance, and the concluding seminar discussion and questionnaires as well as the recorded transcript of the final discussion were translated into English.

The initial discussion was arranged to ensure serious consideration of several possible future developments, followed by a comparative discussion about what could be learnt from each of these possibilities. The seminars lasted around 2½ to 3 hours and consisted of three phases. The participants' task was to discuss what decisions should be made in the near future about NBT.

In the first phase of a convergence seminar, the participants were divided into three *scenario groups* each of which was assigned the task of discussing a scenario of developments up to the year 2020. The topics of the three *scenario groups* focussed on the competitive disadvantages for Europe if the development of NBT were to be severely restricted. The second topic focussed on the potential medical uses of NT together with the accompanying equity issues, side effects, and misuses. The third topic focussed on the use of NBT in diagnostics and surveillance with potential infringements on privacy and freedom of choice.

In *the second phase*, the participants were regrouped into three *convergence groups*, each of which contained representatives from each of the three groups from the first phase. To begin with, they told each other about their respective scenarios and the conclusions drawn in their scenario group. After that, they discussed what could be learnt from comparing the scenarios and what advice they would like to give to decisionmakers. *In the third phase*, all participants met. Each convergence group gave a brief summary of their discussions and recommendations, and then the recommendations were discussed in the whole group. Contrary to the two previous phases, the third phase was audio-recorded. The documentation from the seminars consisted in these recordings and a questionnaire filled in at the end of the third phase by all participants. To illustrate the outcome of the process we present as an example some of the key findings formulated:

1. Facing an emerging and uncertain technology such as NBT, a major consideration is the existing level of trust the public has in institutions, industry, and policymakers. Participants made comparisons with recent experiences with other technological and commercial developments where they felt that the objectives to create a cheap and accessible technology had not been realised.

2. Many important ethical issues are not necessarily the ones that arise in the mid- to long-term stages of the NBT development. The discussions in fact showed that the participants' chief concerns rather lie in the earlier stages of the development, for instance in deciding on research priorities and regulation.

3. Further public participation on social and ethical issues of NBT was unanimously encouraged. A broader public influence over the technological development was supported in its own right, but it was also advocated as a necessary step in avoiding (additional) public alienation or backlashes and in enabling NBT to be beneficial to different members of society.

4. Views were divided on how much regulation is needed to curb unwanted developments. Some participants believed that early regulation is necessary as it deters unwanted applications. Others argued that regulation (and bureaucracy) is important as it at least delays the development, thereby allowing for more insight into long-term impacts and side effects. In contrast, some said that over-regulation and

excessive precaution were a problem because it implied a loss of potential benefits and would also generate (economic) unbalance if certain countries went ahead with less regulation. A common feature in all responses was the recommendation to make the regulatory bodies robust to meet the demands of novel technologies.

5. As a methodology for facilitating a forum for decision-making under uncertainty the Convergence Seminars were by and large successful. The responses the participants gave in the discussions and questionnaires indicated that their advice on what decisions should be made about the NBT development was influenced both by different possible future developments and also the point of view of different individuals.

10.2.2   The DEMOCS Card Game

The purpose of the DEMOCS Card Game on applications of NBT was to engage with ordinary citizens without special knowledge of the subject in their own local context and to test this out by playing the game in a variety of groups. The initiator of the process was the Nanobio-RAISE Research Consortium,[5] in collaboration with the New Economics Foundation (nef),[6] who devised the DEMOCS concept in 2002.[7] DEMOCS is a novel form of lay participation in the format of a card game, around which participants discuss for 2 hours and come to agreed or divergent views on specific policy issues or on general principles. In the type of engagement and the information it can provide, this card game method comes between opinion polling and focus groups.

The process was funded by the FP6 supported Nanobio-RAISE project (€553,845).[8] The DEMOCS Card Game was carried out under the responsibility of the Society, Religion and Technology

---

[5]See more about the Nanobio-RAISE project on http://files.nanobio-raise.org.

[6]*nef* is an independent 'think-and-do' tank (http://www.neweconomics.org/gen/).

[7]The game was originally devised with sponsorship from the Welcome Trust by Perry Walker of *nef*, in consultation with experts in ethical and social issues and in innovative engagement with publics. The first game was on stem cells and cloning. Different versions have been successfully devised and used, for example, as one of the methods used in the UK GM Nation Debate on genetically modified (GM) crops in 2003, in genetic testing kits for the Human Genetics Commission, and diverse issues such as climate change or animal research.

[8]See more about the project on http://files.nanobio-raise.org.

Project (SRT),[9] a consortium partner of the Nanobio-RAISE project until 6th July 2007. Since this date the work has been carried out by Edinethics Ltd.[10] as successor of SRT as a project's partner. The director of Edinethics Ltd., Dr Donald Bruce, was the coordinator of the DEMOCS Card Game onNBT. He has experience with deliberative methods, so this was not his first experiment in this field. The cards were written by Donald Bruce and Perry Walker from *nef*. Some of these cards were taken from the existing *nef* game on broader issues of NTs, but many cards were written specifically for this game. The outcomes of the Convergence Seminars gave support in framing some of the cards.[11]

The first card game on NBT was played by participants at the Nanobio-RAISE Oxford course on March 2007. Following this, the game was revised, and a final version was available from June 2007. The main topics of these deliberative processes were five medical applications for using NT: to assist in early diagnosis, to remotely monitor the inside of a body, to target drugs to disease sites in the body, to regenerate cells and tissues, and potentially to electrically stimulate brain by nanodevices without major surgery and two non-medical applications, namely to add nanoparticles to fortify foods with omega fatty acids and to use converging technologies for human enhancement. Consumer goods were concerned with the non-medical application of adding nanoparticles to food.

The basic questions to be discussed were introduced by *Story Cards*, which had imaginary cases that illustrated issues and dilemmas based on credible or real-life examples. In this game there were seven Story Cards in which seven applications of NBT were given, and at the end of the game the players were invited to vote for the acceptability of these applications. The players read these stories to each other as a group and discussed them briefly. This was followed by a second round based on *Information Cards*, which provided some of the information which is needed to understand what the various technologies entail. The third round follows the same process with the *Issue Cards*. These cards give a range of different perspectives,

---

[9]SRT is a unique unit of the Church of Scotland set up in 1970 to examine some of the vital issues of our times. SRT has been much involved in the recent UK public debate on GM crops (http://www.srtp.org.uk/).

[10]Edinethics Ltd. is an independent consultancy set up in May 2007 to provide technically informed and balanced assessments of ethical and social issues in current and future technologies (http://www.edinethics.co.uk/index.htm).

[11]See http://files.nanobio-raise.org/Downloads/nbrdemocs.pdf.

questions, and viewpoints about issues raised by the technologies and are meant to provoke debate. During this process, themes begin to emerge. Participants are encouraged to work these into one or more opinion statements written on the *Cluster Cards*, listing the different cards they have found useful in forming this statement. Ideally these would be consensus views of the whole group but they may also be the views of individual members. This is entirely up to the group. *Blank Cards* are also available if players want to add any points or questions originally not included in the cards. At the end of a card game, participants are asked to vote on a series of policy options. In this case they were invited to give their view on the acceptability of seven different applications of NBT given in the seven Story Cards discussed at the beginning.

The game lasts between 1 and 1½ hours. The participants meet only one time, when the actual game is played. About 20 games were played during the period June to August 2007 in seven locations in the UK and the Netherlands. The NBT DEMOCS Card Game was played further on until October 2008.[12]

The participants in the DEMOCS Card Game on NBT were non-specialists. Not all relevant stakeholders were involved in the game, since this was not the aim of the method. The facilitators were selected mostly from people with previous experience of DEMOCS Card Game. Facilitators were prepared to gather groups of people to come together to play and to act as game masters. At each location, the local facilitator welcomed participants and explained the aims of the game, the process to be followed, the outcomes, and importantly, the ground rules for good discussion. All participants were engaged in the semi-structured group discussion and at the end of the process all participants were invited to vote for the acceptability of the selected seven applications and to formulate in their own words the values that were important to them in coming to their views. The groups were from a variety of contexts and included MSc students in Amsterdam, a gay and lesbian group in Manchester, and thirty-two 15-year-old school pupils in Hertfordshire. They involved over 130 people of whom 114 recorded their votes on the options.

---

[12]See Dr Donald Bruce: Engaging citizens on nanobiotechnology using the DEMOCS game. Interim Report on DEMOCS games on nanobiotechnology played in the UK and the Netherlands, 2007.

10.2.3 The Nanologue Project Dialogue

The Nanologue project was funded by the EC's 6th Framework Programme and initiated by a research consortium formed by four partners. The budget provided by the EC to realise the Nanologue project accounted for €339,663. The coordinator of the project was the Germany-based Wuppertal Institute for Climate, Environment and Energy, which has a focus on application-oriented sustainability research and addresses the major challenges related to sustainable development, such as climate change or resource shortages. The institute has an interdisciplinary approach and works towards systems understanding.[13] Among the partners we find the Switzerland-based EMPA, the ETH (Swiss Federal Institutes of Technology) domain's institution, as an interdisciplinary research and services institution for material sciences and technology development. The focal points of EMPA's research work are arranged in five programs with the following themes: nanotechnology, adaptive material systems, materials for health and performance, natural resources and pollutants, and materials for energy technologies.[14] The UK-based Forum for the Future, which is a sustainable development charity,[15] and the Pan-European applied research and training organization called Triple Innova also took part in the management of the project.[16]

The purpose of the Nanologue project was to initiate a dialogue on NT that brings together a wide range of technological experts and civil society representatives (Nanologue, 2006b). The dialogue processes were used to provide a neutral platform gathering recent knowledge and evidence about the multiplicity of potential impacts associated with selected NTs. As an overarching objective, the project attempted to facilitate a dialogue among various stakeholders about the benefits and potential impacts of NT. The project objectives according to its three phases were (1) mapping the ELSA of NTs, (2) consulting with representatives from research, business, and civil society, and (3) constructing tools to enhance further dialogue on NTs. The consultation process attempted to engage with representatives from research and development as

[13] www.wupperinst.org
[14] http://www.empa.ch
[15] www.forumforthefuture.org.uk
[16] www.triple-innova.de

well as marketers, users, and accommodators of NT in a dialogue in order to capture their perspective on the issue. Within this consultation process they interviewed the societal groups engaged in this project phase in order to further develop and substantiate the benefits and potential impacts of NT (identified in the first project phase). One of the objectives of the consultation process was to facilitate the process of prioritizing the benefits and potential impacts identified. The consultation process also contributed to the development of scenarios exploring how business, civil society, and public institutions can engage effectively in a dialogue on the ethical, social, and legal aspects of NT and a web-based tool to quickly assess possible ethical, legal, and social issues for NT products and research proposals.

The three-phase project continued for 21 months; it started on 1st February 2005 and finished on 31st October 2006. There was continuous consultation with leading experts of the field, who functioned as an external advisory board to the project, and in the second phase, the consultation was extended to other experts and civil society representatives for a 8 months, from September 2005 to May 2006. Within this period experts were telephone interviews were and workshops have been visited for interviewing experts. For a dialogue with civil society, participants were invited to a one-day workshop on the 5th September 2005.

The project intended to follow an iterative approach in which restrictions narrowed down the project's scope. The selected NT application areas were energy conversion and storage, food packaging, and medical diagnostics. The selected ELSA topics were environmental performance, human health, privacy, access, acceptance, liability, regulation, and control.

The different participants in the project were the research consortium as the initiator of the process, selected researchers, civil society representatives, and the external advisory board. Selected researchers were scientists/technologists involved in research and development of NT-based applications, both in university and business. They were working on the following three NT application areas: medical diagnosis, food packaging, and energy conversion and production. In addition, one group was invited to participate in the interviews disregarding their specific NT area of expertise. This "generic group" has been interviewed on ELSA and NTs in general. Several researchers contacted by the Nanologue team were not able to participate. Usually they reported to be either overloaded with interview requests or were simply not interested in participation. Thus, the conclusion can be drawn that the group of experts who agreed to participate in the interviews takes an above average interest in NT-related ELSA. Some explicitly named the importance of the Nanologue project aims as the primary reason for their participation (Nanologue, 2006a). Civil society is usually defined as the totality of voluntary civic and social organizations or institutions. However, for the purpose of this project the group was widened to include those who were involved or interested in understanding, marketing, regulating, monitoring, and writing about NTs or were helping to develop market-relevant products using NT applications. As a result this group was expected to have a better understanding of the issues surrounding NT than a more traditional group of representatives from civil society.

The involvement of civil society representatives was not meant as a representation of the public's perception of NT but to ensure balanced views from a wider group than just researchers. Actually, they had a rather limited role in the project's second phase, that is, to participate in a one-day workshop. The workshop was used to further explore the ethical, legal, and social aspects of NT and to develop recommendations for communication of ethical, social, and legal issues between different actors in EU in the future. During the workshop the delegates were asked to list the main risks and benefits of NTs, vote for the most important risks and benefits and prioritise them, vote on how certain they were that these risks and benefits would be relevant by 2015, discuss the barriers to better communication between stakeholders concentrating on the role of scientists, and make recommendations to overcome these barriers. The results from the interviews with researchers and the workshop with civil society were analysed and compared. The external advisory board consisted of 9 members with different academic backgrounds, but all with expert knowledge on various aspects of NT, covering both the techno-scientific and the ELSA of NT. They had a continuous role throughout the project as they complemented the consortium of expert knowledge, critically review project findings, and support outreach activities of the project. The project's general aim was not to represent the common voice but to prepare the ground for public dialogue. In the second opinion phase of the project, stakeholders were invited to participate in the dialogue, but most of them were experts and representatives of research or industry with little lay public involvement.

10.3 Review of the Processes

The role of deliberative processes in shaping the nanotech debates and governance has been critically discussed in many scientific papers already.[17] However, new insights can be provided based on empirical approaches of selected contemporary experiences of deliberative processes that took place in various contexts (Hagendijk & Irwin, 2006:168). In this chapter we would like to restrict our analysis to the evaluation of three reviewed deliberative processes' outcomes in relation with their proposed objectives as well as with the EU strategy in this field, since the processes we analysed gain their relevance and meaning in this context.

As we presented in the introduction, the EC's Strategy and Action Plan on Nanotechnology and AP on SAS emphasise the importance of involving/engaging a large number of stakeholders in the debates about NT and taking into account the societal and ethical aspects from the earliest possible stage of technology development. The EC funded a number of projects that aimed to contribute to the achievement of these objectives.[18] The processes reviewed here aimed to shape the debates on the social and ethical aspects of NT and to develop and use various methods to enhance such dialogue. Thus, the novelty in the majority of the processes we overviewed was not only the topic but the method of deliberation itself. The Convergence Seminars model of engagement developed at the Royal Institute of Technology, Stockholm, to facilitate discussion and decision-making about emerging technologies was first applied in practice in the Nanobio-RAISE project. Even though the DEMOCS Card Game method was developed in another context, its application in the field of NBT was developed for the Nanobio-RAISE project. The aim of these initiatives was to add value to the existing state of the art by applying these methods to discuss issues raised by NT applications and to experiment with how these methods work in different contexts. They shall thus be evaluated against these

---

[17]Kearnes and Rip (2009), Macnaghten *et al.* (2005).

[18]For an overview of the funding activity of the EU in the field of nanotechnology, see EC (Dr Angela Hullman, 2008).

objectives. What is therefore interesting in this review is how these methods could facilitate the dialogue on NT, to what extent they could involve participants, how broad was the circle of citizens who were involved, how cost and time effective are the processes, and how various contexts impact upon the results of the processes. In addition to these criteria we will also assess the reviewed processes according to Cohen's four criteria of deliberative process (Cohen, 1989).

10.3.1   Initiatives and Objectives

The Nanobio-RAISE project took place when EU policies on the specific area discussed were still about to be shaped. The project aimed to anticipate the societal and ethical issues likely to arise as NBTs develop and to use the lessons from the European GM debate to respond pro-actively and responsibly to the probable public, media, and political concerns. It aimed to try out new ways to involve the public in discussions on nanoethics by using the convergence seminar method and a novel form of lay participation based on DEMOCS Card Game. The stated objectives of the Convergence Seminars were rather general, and experimentation with public deliberation was not explicitly mentioned. However, one may conclude that to improve the methodology used by the Convergence Seminars was another implicit purpose of the process, and the project fulfilled this objective. The DEMOCS Card Game method was used in the project to map citizens' views and expectations related to the topic of NBT. The DEMOCS Card Game method enabled a much wider spectrum of the population to engage with new issues than what would have been possible with most qualitative public engagement methods. The findings of these kinds of studies, while not statistically quantifiable like those of large population surveys such as the Eurobarometer, may provide much greater insight into the underlying values, motivations, and desires involved and so inform policymakers about public's views.[19]

The Nanologue project was launched to bring a wide network of experts and some civil participants into a focused dialogue on the ELSA of NT in an early phase of technology development.

---

[19]See more in Nanobio-RAISE: Public Perceptions and Communications about Nanobiotechnology (http://files.nanobio-raise.org/Downloads/NanoPublicFINAL. pdf).

The project intended to facilitate further dialogue and public participation activities on NT in Europe. At the time of the research and consultation process, the societal environment already grappled with the new technology in numerous ways. However, most approaches into the ELSA of NT have been initiated by academic and research institutions that could be contrasted with the paucity of civil society activities trying to influence the development of this new technology. The period could also be characterised by the fact that the public had a scarce knowledge about the technology and its possible impacts on society. The process organised in the framework of the project thus contributed to raise awareness about these topics and the insights gained throughout the project were translated into a selected number of recommendations that are expected to facilitate the dialogue on NTs and ultimately a better integration of ELSA into NT development.

10.3.2 Organisation

Since both the Convergence Seminars and the DEMOCS Card Games[20] were part of the larger research project, Nanobio-RAISE, it was difficult for us to gain reliable and precise data on the monetary costs for organizing the procedure. In order to realise such a process, the monetary costs can be calculated by estimating the personal costs for the moderators and the organiser, the travel expenses of the participants, and the catering costs and rent for the meeting rooms. As there were relatively low number of participants and contributors for each event, and the events were located in universities and most of the participants were scholars or students, we can assume that the monetary costs were reasonable. Based on the information gathered, we can state that both processes organised in the framework of Nanobio-RAISE project were equipped with an adequate amount of resources. The scale of the process was appropriate to the purpose and objectives of the project. In the case of the Nanologue project we estimated that a significant proportion of the budget covered personal costs of the numerous experts involved in the 21-month-long project and a smaller amount might be spent on some additional costs, such as travel and accommodation costs

---

[20]According to Involve's assessment, the development of a game costs from £10,000 to £20,000 and upwards; see http://www.involve.org.uk/mt/archives/blog_37/Democratic%20Technologies.pdf.

of the meetings and workshops. The fact that the project has been successfully finished might lead us to conclude that it was equipped with a reasonable amount of resources. The time allocated to the processes, their duration, and their frequency seem to be enough to address the issues to be discussed.

In case of all three processes, experts from various fields were invited to cover the content issues related to NT, and in addition to this, persons with previous experience in conducting such deliberative processes were used as moderators or advisers.

10.3.3 Participation

Summarizing the participatory aspects of the Convergence Seminars, we can conclude that there were a few (1–4) participants with background in technology or natural science. In terms of age and gender the process achieved a fairly good balance between the seminar participants, with the exception of Visby and Porto where the majority were women. In terms of socio-cultural background, students were an over-represented group at all the seminars. According to the organisers' explanation, this might be due to their interest in the topic and to the academic setting that hosted the seminars. Participants represented a geographic diversity of Europe that was accomplished by hosting the seminars in four geographically distinct regions in Europe. The selection of participants also aimed towards including people of different age, gender, and social and cultural background at each location.

However, as the final report of the Convergence Seminars highlights, the size and distribution of the invited participants limit the process capacity to be used as a representation of the common voice. The scope and the content of the process was clearly defined and explicitly declared to the participants. There was some degree of thematic restrictions, the discussions being focused on a number of selected and clearly defined topics guided by three scenarios.

In the case of DEMOCS Card Game the participants were non-specialists. The facilitators had been selected mostly from people with previous experience of DEMOCS Card Game. The local facilitator welcomed the participants and explained the aims of the game, the process to be followed, the outcomes, and importantly, the ground rules for a good discussion. The groups were from a variety of contexts and over 130 people were involved in playing the card game. Only 114 of them recorded their votes on the options.

The Nanologue project attempted to facilitate a dialogue among various stakeholders about the benefits and potential impacts of NT. The different participants in the project were the research consortium as the initiator of the process, selected researchers, civil society representatives, and the external advisory board. Selected researchers were scientists/technologists involved in research and development of NT-based applications, both in university and business. The representation of the civil society was very limited.

10.3.4 Results of the Processes

The outcomes of the process we could evaluate were in most cases the recommendations formulated by policymakers (Convergence Seminars, Nanologue) or votes on certain NT applications (DEMOCS Card Game). To illustrate outcomes of one of the processes of the Convergence Seminars, we present the list of recommendations produced by the participants:

1. Decide on research priorities that target crucial needs in society. For instance, socially useful applications that target global environmental problems and benefit human health should be prioritised over novel consumer products and military applications.
2. Research should generate applications that truly benefit the developing world, thereby contributing to decreasing rather than increasing the gap between countries.
3. Research should be directed towards finding solutions for the less privileged part of the population within wealthy countries.
4. Create a variety of forums for debate and discussion of NBT in order to promote curiosity about NBT in the public.
5. Encourage different actors (researchers, policymakers, industry, and media) to take responsibility of public engagement at all stages of the development and ensure that the public receives unbiased information.
6. Develop programmes for international cooperation across both research and regulation of NBT. Focus the international cooperation on mutually important problems, such as combating global climate change.

7. Involve a range of researchers from different scientific disciplines in the NBT development. Do not conceive of NBT as a strictly technological or industrial venture.

8. Prepare for a debate on arising controversial issues of privacy, freedom of choice, and enhancement and distributive justice in medical applications of NBT.

All the medical applications had broad approval with at least half voting "yes" and a large majority "yes" or "possibly". In contrast, human enhancement was viewed much more negatively, with around half voting "no" or "doubtful", and a relatively small number voting "yes". Current fears regarding NBT were dominated by the issue of nanofood technology, commonly associated with GM foods. Also dominating the discussion were certain ideas rooted in science fiction such as replicators. Nanomedicine was viewed in a much better light, with certain advancements being heralded as great achievements. These include improved diagnosis, treatment, and monitoring of patients, particularly in areas such as cancer and cardiovascular and neurodegenerative disease.

It has to be mentioned that not only recommendations emerged as a result of these processes (Convergence Seminars and DEMOCS Card Game), but a few common lessons were also formulated:

1. Public attitudes are formed not only in relation to particular technologies but also to the policies and values that influence the direction of technological development and to the social and political conditions in which they develop. People are not only concerned about the potential benefits and risks of NTs but also about who is most likely to be affected by the benefits and risks.

2. Public attitudes towards risk, uncertainty, and regulation tend to be concerned with their views of the ability of regulations and regulatory authorities to manage complex risks, whether hypothetical or actual. This is notwithstanding whether regulation is really necessary or can be framed in anticipation. Although many concerns focus on potential risks such as toxicity of manufactured nanoparticles, members of the public seem also to be concerned with government's and industry's ability to deal with potential long-term risks and uncertainties associated with NTs as with the risks themselves. This includes a concern among some about the

government's and industry's ability to ensure that potential benefits and risks are distributed in ways seen to be equitable.

3. The issues raised by the members of the public relate mostly to broad aspirations and concerns about future implications of NTs rather than particular technological developments. Even when more specific issues have been the focus of discussions, the final recommendations have tended to be broad in scope, addressing topics such as "all manufactured nanoparticles" or "companies using NTs in the environment".

4. There is a consistent demand for more open discussion and public involvement in policymaking relating to science and technology overall than has been afforded up to now.[21]

The Nanologue project had a deliberative and participatory character; however, it was not a deliberative process *per se* but an important step towards future initiatives and provided easily usable tools (Nanometer and the three scenarios) that support exercises in public participation. Thus, the project gave a structured representation of expert opinions on the ELSA of NT focusing on specific near-term applications. Recommendations made by the Nanologue project on future dialogues on NT involved the following:

Frame the dialogue: NT "vision" should be refocused, not around the risks but around a vision of sustainable development, which can be used as a framework to maximise benefits and minimise risks. However, caution must be placed not to fuel a nano-hype that might lead to a nano-bubble.

A dialogue in context: It should enable to differentiate between the sometimes profound differences between a wide array of applications currently captured by the term "nanotechnology". Contextualisation in terms of need for certain applications is important. Contextualisation also enables one to focus on specific applications instead of NT in general that should be followed in the construction of regulatory frameworks.

An open dialogue: The process must be as open as possible but should consider the fact that too much information can lead to confusion or disengagement.

---

[21]These results have been summarised in Nanobio-RAISE: Public Perceptions and Communications About Nanobiotechnology (http://files.nanobio-raise.org/ Downloads/NanoPublicFINAL.pdf).

Assessing the risk: Risks of nanoparticles and other nano-based applications' to human health and environment are still not clear. International standardisation would serve a basis to enhance risk assessments. More consistent and integrated approach is needed to achieve sound risk assessment.

Making information accessible: Transparency in itself is not enough if one cannot get access to the information easily enough to be involved. Beside conferences being a prime source of information in the case of researchers, the internet is an obvious candidate for effectively providing easy accessibility. However, it is true that the internet provides a large amount of information about the ELSA of NT but in a scattered way.

Possible solutions towards these goals were also formulated. Such solutions were to set up a "clearing centre" hosted by an institution with high social legitimacy to initiate activities that involve the general public via museums or science centres, to engage producers of products containing NT components with retailers in labelling NT products (some participants were cautious about this solution), and to target journalists and the media with information both about benefits and risks of NT applications.

There were certainly other "results" that cannot be translated into recommendation or "common lessons" and assessed through desk research. These include the impact of these processes on the participants, the personal experiences gained through participation, the new insights that the debates and discussion provided for the participants, and possible frustration related to participation, to the outcomes, and to the impact of the processes.

10.4 Deliberation Criteria

According to Cohen the following criteria can be used to evaluate deliberative processes: is it a *free discourse*? Are there any thematic restrictions, and if so, how are they explained or legitimised? Is it a *reasoned process*? Are the *participants equal*? Is it a *consensus-driven* process? What forms of interaction between participants, for example, working groups, hearings, online communication, etc., are employed in the process? Does the process make use of scenario techniques? How are scientific and other reasoned arguments brought into the process? Who provides these inputs? Is there a variety of opinions included in the process and if so, in what way? Is there room for mutual learning and change in positions? Can the process be regarded as transparent and if so, in what terms? (Cohen, 1989).

The Convergence Seminars employed "classical" means of interaction and face-to-face communication in a structured way. Scenario techniques were used as described above. Scientific arguments were brought into the process through the usage of scenarios. The process intended to develop the possibility of mutual learning between the participants and so moved the participants from their initial group's scenario to another "mixed" scenario group where initially developed perspectives and insights were open to challenge. In the third phase participants gathered together to summarise their insights and formulate their advice, which resulted in a list of recommendations. Although the three phases of the process lead to a consensus between the participants, this was not a prerequisite for the process. According to the initiators of the seminars, "the term 'convergence' thus refers to the converging structure of this seminar model — not necessarily to convergence of opinion amongst the participants".

The Convergence Seminars can be regarded as successful since they provoked discussion, and according to the organisers, the participants gave many positive feedbacks. We lack information on the exact costs of realising such a process, but we could still anticipate that it is a cost-effective way of involving the public in a deliberative process. Although the main objective of the process was to explore the public views on the development of NBT, the organisers emphasised that considering the small sample size of the participants in the seminars the comments should not be interpreted as representative of the common voice of the specific country. However, in the organisers' view the advice given at the seminars by the participants can be taken as an advice given by a reasonable variety of public in Europe.

The form of interaction between the participants was face-to-face discussion while playing DEMOCS Card Game. The process makes use of scenario techniques in the form of Story Cards. Scientific arguments were brought into the process through the Information and Issue Cards. These inputs came from the organisers of the game. There was room to express opinions, but the possible opinions were to a certain degree framed by the cards provided by the organisers.

However, the participants had the opportunity to introduce new issues through the Blank Cards. There was room for mutual learning during the discussions. To a certain degree the DEMOCS Card Game can be regarded as a transparent process. According to some analysts such meetings tend to attract almost exclusively those with already committed views in relation to the topic and quite possibly polarise them further in their pre-determined opinions, negative or positive, as a result of the experience. In the case of DEMOCS Card Game people without previous knowledge about this technology were reached in their own locations so that the spectrum of the population reached could at least be widened by this method.

In the case of the Convergence Seminars, there was a close to complete agreement among the participants when asked what they wished to advise the policymakers deciding on the development of NBT. In the DEMOCS Card Game the aim was not necessarily to arrive at a consensus, and each participant voted individually at the end of the game. If there was some sort of consensus at the end, it was not a "forced" one. In the case of the DEMOCS Card Game we were evaluating, quite a high level of consensus was achieved.

The strength of the methods is that the game enables participants to engage with these issues in their own environments and locations and also provides a degree of empirical data. Its advantages over conventional focus groups are that it can be played by any group of people who care to take part, anywhere, and with much less reliance on expert facilitation. It can provide considered views and comments from a much wider range than are tapped by normal focus group sampling and far more depth than from tick box opinion polls. It also has a multiplicative value in that, once devised, the game can be played in any context anywhere by any group of people. Although devised for use as hard copy with known groups, web versions of many DEMOCS Card Games are also now available for a wide audience. The weakness of the method is that the results are loosely connected with concrete policy outcomes. Also the questions that to what extent can we extrapolate the data from these groups and how representative are the outcome?

The Nanologue project, in all its phases, restricted the involvement of a wide array of public participants and stakeholders and successfully engaged with experts in ELSA of NT and with researchers in the NT field. As it has been already mentioned, the project comprised three interrelated work phases: mapping the literature, collecting opinions through consultation, and building three scenarios. Scientific arguments were fed into the process at all three phases from the external advisory board and the consulted researchers. While civil participation was rather limited, mutual learning and exchange of views could be realised between the representatives of different expertise. Thus, the variety of opinions suffers from the lack of real public voices. However, regarding the early stage of technology development and the general goal of the project to facilitate further dialogue and create tools that might enhance public participation, this weakness does not seriously hamper the project legitimacy. Scientific evidence and a variety of expert opinions are important requirements in the future development of deliberative processes on NT. The transparency of the process was adequate, with media coverage and a webpage that made the project content easily accessible. The limitation of the topics discussed on energy, food, and medical diagnostics could be represented as an advantage as it enabled a focused and efficient dialogue on the issues.

While summarising the strengths and weaknesses of the project, we considered the fact that the initiators of the process understood their activity not as a classic deliberative process but more as a preparatory exercise that could serve as a significant facilitator in deliberative processes on NT. As a participatory and deliberative process, its strength lies in its argumentative and structured dialogic nature, although in its well-defined scope and its high level of transparency that characterise the process. The general weakness of the process comes from its limited participatory framework that predominantly focuses on expert opinions. The predominant expert focus of a process might generate a "framing problem" with silencing or pushing to the periphery those public voices that cannot be fitted into the specific ELSA of NT framework created mainly by consultation with experts and with researchers. The results of the process can give substantive support in the planning and realisation of public deliberation on the ELSA of NT in specific application areas.

10.5 Summarizing Appraisal

All the three processes that we have evaluated can be best understood as an experimentation with deliberative processes. Therefore their primary value is not their impact on the field of NT policy, but the development and experimentation with deliberative methods through which the problems raised by the various technologies can be discussed and debated.

The processes we reviewed can be also relevant in demonstrating the fact that if the "public" is provided context and information, it can be meaningfully involved even in debates on new technologies such as NT. On the basis of the outcomes of these processes and in line with the conclusion drawn from the evaluation of other deliberative processes conducted in other contexts, arguments that the "public" cannot meaningfully contribute to the debates and discussions can be, at least, seriously questioned. However, further efforts should be made to assess which method works better and more cost effective in which context[22] and how to promote more fruitful and effective interaction among various stakeholders (Jasanoff 2003:238).

While the impact of these initiatives on member states NT policy is difficult to assess, there are very good chances that the result of these processes and initiatives will create a "deliberative background" for EC's NT policy on which they might have even some direct impact. Considering the fact that in some important areas the EU level policy will precede those of member states (e.g., in the area of novel food) this possible impact is not a negligible one.

The development of such processes shall not stop in this phase. Once methods developed through initiatives like those presented in this chapter are available and prove to be effective in debating NT with a larger circle of stakeholders, it is very important to consider how the "deliberative experiments" and experiences can be integrated in the fabric of "real" policymaking both in the EU and in the member states. Recommendations were already formulated with this regard in a number of published reports and some steps in this direction have already been implemented in the "new" generation of deliberative attempts funded under the FP7 framework. As participants in one such project, Nanoplat, we can stress the importance of clarifying the goals of deliberative processes, of dealing with the issue of responsibility, and of continuing to increase the involvement of those who can influence the decision-making process. In line with Hagendijk and Irwin's observation (2006), it is important to pay attention to the wider context of governance and to the relationship of public deliberation with more traditional modes of policymaking.

---

[22]See such proposal for example in Gastil (2010).

In agreement with the Nanotechnology Engagement Group (NEC) report, we also consider that too much emphasis on methods carries the risk that other important considerations might be overlooked (Gavelin *et al.*, 2007:138).

References

Official Documents

Commission of the European Communities (CEC) (2009) *Nanosciences and Nanotechnologies: An action plan for Europe 2005–2009*, second implementation report 2007–2009,COM (2009) 607.

European Commission (EC) (2001) *Science and Society Action Plan*, COM (2001) 714, available at http://www.bologna-berlin2003.de/pdf/science_society.pdf

European Commission (EC) (2004) *Towards a European Strategy for Nanotechnology. Luxembourg: Commission of the European Communities*, COM (2004) 338.

European Commission (EC) (2007) *Nanosciences and Nanotechnologies: An action plan for Europe 2005–2009*, first implementation report 2005–2007, COM (2007) 505.

European Commission (EC) (2005) *Nanosciences and nanotechnologies: An action plan for Europe 2005–2009*,COM (2005) 243.

European Commission (EC), DG Research (2008) *European activities in the field of ethical, legal and social aspects (ELSA) and governance of nanotechnology* (Dr. Angela Hullmann), available at ftp://ftp.cordis. europa.eu/pub/nanotechnology/docs/elsa_governance_nano.pdf

European Commission (EC) (2009) Commission recommendation on *A code of conduct for responsible nanosciences and nanotechnologies research & Council conclusions on Responsible nanoscience and nanotechnologies research*, Luxembourg: Office for Official Publications of the European Communities.

Project Publications

Bruce, D (2007) Engaging citizens on nanobiotechnology using the DEMOCS game. Interim Report on DEMOCS games on nanobiotechnology played in the UK and the Netherlands.

Davies S, Macnaghten P, Kearnes M (eds) (2009) *Reconfiguring Responsibility: Lessons for public policy*, (Part 1 of the report on deepening debate on nanotechnology). Durham, Durham University. Available at http://

www.geography.dur.ac.uk/Projects/Portals/88/Publications/ Reconfiguring%20Responsibility%20September%202009.pdf

Gavelin K, Wilson R, Doubleday R (2007) *Democratic technologies? The final report of the Nanotechnology Engagement Group (NEG)*, London, INVOLVE. Available at http://www.nanotechproject.org/process/ assets/files/7060/nano_pen16_final.pdf

Macnaghten P, Davis S, Kearnes M (2010) Narrative and public engagement: Some findings from the DEEPEN project, in *Understanding Public Debate on Nanotechnologies* (René von Schomberg, Sarah Davis, eds), Luxembourg, Office for Official Publications of the European Communities, 14–30.

Montovani, E, Porcari A (2010) A governance platform to secure the responsible development of nanotechnologies: The Framingnano project, in *Understanding Public Debate on Nanotechnologies* (René von Schomberg, Sarah Davis, eds), Luxembourg, Office for Official Publications of the European Communities, 14–30.

Nanologue (2006a) *Opinions on the Ethical, Legal and Social Aspects of Nanotechnologies—Results from a Consultation with Representatives from Research, Business and Civil Society*, London, Nanologue.

Nanologue (2006b) *The Future of Nanotechnology: We Need to Talk*,London, Nanologue.

Rip A, Egan CS (2010) Position and responsibilities in the 'real' world of nanotechnology, in *Understanding Public Debate on Nanotechnologies* (René von Schomberg, Sarah Davis, eds), Luxembourg, Office for Official Publications of the European Communities, 31–38.

Stø E, Scholl G, Jegou F, Strandbakken P (2010) The future of deliberative processes on nanotechnology, in *Understanding Public Debate on Nanotechnologies* (René von Schomberg, Sarah Davis, eds), Luxembourg, Office for Official Publications of the European Communities, 53–80.

van Broekhuizen P, Schwartz A (2010) European trade union and environmental NGO position in the debate on nanotechnologies, in *Understanding Public Debate on Nanotechnologies* (René von Schomberg, Sarah Davis, eds), Luxembourg, Office for Official Publications of the European Communities, 81–108.

Scientific Literature

Besley J, Kramer L, Qingjiang Y, Toumey C (2008) Interpersonal discussion following citizen engagement about nanotechnology: What, if anything, do they say? *Science Communication*, 30(2), 209–235.

Cohen J (1989) Deliberation and democratic legitimacy, in *The Good Polity* (Hamlin A, Pettit P, eds), Oxford, Basil Blackwell.

Fischer F (2000) *Reframing Public Policy: Discursive Politics and Deliberative Practices*, Oxford, Oxford University Press.

Gastil J (2000) Is face-to-face citizen deliberation a luxury or a necessity?, *Political Communication*, 17, 357–361.

Hagendijk R, Irwin A (2006) Public deliberation and governance: Engaging with science and technology in contemporary Europe, *Minerva*, 44(2).

Jasanoff S (2003) Technologies of humility: Citizen participation in governing science, *Minerva*, 41, 223–44.

Kearnes, MB, Rip, A (2009) The emerging governance landscape of nanotechnology, in *Jenseits von Regulierung: Zum politischen Umgang mit der Nanotechnologie* (Gammel S, Lösch A, Nordmann A, eds), Berlin, Akademische Verlagsgesellschaft.

Kearnes M, Wynne B (2007) On nanotechnology and ambivalence: The politics of enthusiasm, *Nanoethics*, 1(2).

Laurent B(2008)Engaging the public in nanotechnology? Three visions of public engagement, Centre de Sociologie de l'Innovation Ecole des Mines de Paris (working paper).

Macnaghten P, Kearnes M, Wynne B (2005) Nanotechnology, governance, and public deliberation: What role for the social sciences? *Science Communication*, 27(2), 268–291.

Pidgeon N, Rogers-Hayden (2007) Opening up nanotechnology dialogue with the publics: Risk communication or 'upstream engagement? *Health, Risk & Society*, 9(2), 191–210.

Powell M, Kleinman DL, (2008) Building citizen capacities for participation in nanotechnology decision-making: The democratic virtues of the Consensus Conference model, *Public Understanding of Science*, 17(3), 329–348.

WebPages

CORDIS: Nanotechnology Research at the European Commission. http://cordis.europa.eu/nanotechnology and http://cordis.europa.eu/fetch?CALLER=FP6_PROJ&USR_SORT=EN_QVD+CHAR+DESC&QZ_WEBSRCH=nanotechnology&QM_EP_PGA_A=FP6-SOCIETY&QM_EP_CT_D=&QM_EP_CY_D=

European Commission. Science in Society portal. http://ec.europa.eu/research/science-society/index.cfm?fuseaction=public.topic&id=1221

Nanotechnology at the European Commission: http://ec.europa.eu/nanotechnology

Dem Net: http://search.conduit.com/Results.aspx?q=deliberative+process+Europe+science+and+technology++&SearchSourceOrigin=3&gil=en-US&hl=en&SelfSearch=1&ctid=CT697312&octid=CT697312&start=30

EUROPA Industrial Research homepage: http://ec.europa.eu/research/industrial_technologies/index_en.cfm

Chapter 11

Standardisation as a Form of Deliberation

Harald Throne-Holst and Pål Strandbakken
SIFO Consumption Research Norway, Oslo Metropolitan University,
Postboks 4, St. Olavs plass 0139 Oslo, Norway
harth@oslomet.no

11.1 Introduction

The aim of this chapter is to discuss to what extent one could claim that standardisation is a deliberative process, with a particular focus of the standardisation work on nanotechnologies that takes place in CEN and ISO. For this purpose we will present some general information on this particular work, and not a comprehensive overview of the activities. This would indeed demand more than a mere chapter, and more of a book itself: There are currently (as of November 2012) 234 documents concerning the work in CEN available in the technical committee (TC) 352. For the activities in the corresponding committee in ISO, ISO/TC 229, more than 1000 documents (1033) are available. A presentation based on that

Consumers and Nanotechnology: Deliberative Processes and Methodologies
Edited by Pål Strandbakken, Gerd Scholl, and Eivind Stø
Copyright © 2021 Jenny Stanford Publishing Pte. Ltd.
ISBN 978-981-4877-61-9 (Hardcover), 978-1-003-15985-8 (eBook)
www.jennystanford.com

amount of information would be a demanding task. And, it is in our opinion not necessary to evaluate the deliberative aspects of these activities.

A standard is a technical document that is used as a rule, guideline, or definition. Essentially, it is a consensus-built, repeatable way of doing something. Standards are created by bringing together all interested parties such as manufacturers, consumers, and regulators of a particular material, product, process, or service. Standards can also be used to describe a measurement or test method or to establish a common terminology within a specific sector.

Standardisation work formally aims at producing standards. But it may have other effects as well. It brings together important stakeholders in the same room. During the meetings they interact and exchange viewpoints and opinions. For an emerging technology field like nanotechnology this can be specifically important. In such processes visions of the state of the art of the technology as well as common visions of future developments are produced (Delemarle & Throne-Holst, 2010). This type of visionary work among stakeholders helps to bring some kind of coherence to a diversified and perpetually changing field.

In the description we will focus on the European standardisation activities under CEN (The European Committee for Standardization), but at relevant points we also refer to the International Organization for Standardization (ISO), or global activities in the field. Under the Vienna agreement, the standards community seek to avoid overlapping work and tricky jurisdictions, and empirically it seems as if ISO presently is the most active body in the nano field: 'For topics of mutual interest to ISO and CEN, it is expected that work should be carried out under the Vienna Agreement with ISO lead.'[1] This was further elaborated in a mandate issued by the European Commission in 2010 (EC, 2010): 'Priority should be given to the work carried out in conjunction with ISO'.

Unlike the previous presentations, this chapter does not necessarily deal only with a specific deliberative process, firmly placed in time and space, but rather with a mixture between an ideal type of standardisation process in the nano field and some early experiences with the development of anticipatory standards for it. One possible perspective on standardisation work that we like to mention without going deep into it is that it seems to be related to

---

[1]http://www.cen.eu/cen(Sector/Sector/Nanotechnogies/Pages/default.aspx

what actor–network theory (ANT) authors write about 'stabilisation' of impulses and technologies (Bijker & Law, 1992).

The importance of standardisation work for an emerging technology like nano is known and acknowledged: 'The vital role of standards in the emerging debate on nanotechnology risk assessment cannot be overestimated', and 'Standardisation has clearly been identified as one of the most important issues in the context of a responsible development of nanotechnologies, and many respondents have mentioned the key role of standardisation as a parallel process to the development of definitions, metrologies and methodologies' (Framing Nano Final Report 2010, pp. 97–98, in Forsberg 2010).

The main point of interest here, however, is not standardisation as such, but rather the participatory and deliberative aspects of such processes. How do the European (CEN) and international (ISO) standards bodies define, use, and regulate the deliberative or democratic component in their effort to standardise nanoscience and nanotechnology?

Most of the standards produced by both the CEN and ISO are technical standards for products, however the standardisation activities on services are on the rise (CEN 2012), (EC 2012). The products standards describe technical specifications that should be met before a producer can claim to meet the standard in question. In the technical committee for Nanotechnology in ISO, the ISO/TC 229, it is joint working group 1 that elaborates the standards for terms, definition, and vocabulary for nanotechnologies. Among the issues are the definitions of what nanotechnology and nanoparticles are. In principle one would think that this is a matter of technical specification. But it turns out that there is disagreement on where the endpoints for what constitutes a nanoparticle are: is it between 1 and 100 nm exactly, approximately, or should it be extended to 300 nm or even 1000 nm. These disagreements seem to at least partly have their origin in values; in the sense that in the face of the uncertainties that surround nanoparticles and nanotechnologies, some actors, like environmental organisations, would like to have wider definitions, ,whereas others find that a wider definition may be unnecessary caution and will include particles, agglomerates or aggregates that may not exhibit novel properties. A legitimate question is whether such value-laden questions could and should be solved in a working group with the current scope and framing.

Is it fruitful to act out conflicts of interests as disagreement over definitions?

11.2 Description of the Process

Standardisation can be said to operate at three levels: globally through ISO, regionally like Europe and CEN, and nationally, like Standards Norway.[2]

In general, standards are supposed to contribute by attending to demands to health, environment, and safety. They make the development, production, and supply of products and services more efficient, safer, and cleaner. In addition they aim at making bilateral and multilateral trade between countries easier (ISO, 2008).

The relation between ISO and CEN is regulated through the Vienna agreement. This agreement seeks to limit double work and prevent duplication and increase the transparency between the two. It is an agreement on a technical cooperation between the two.

In 1985 the European Council adopted *the new approach to harmonisation and standardisation*, often referred to just as 'the new approach'. Under this, the Council defines *essential requirements* especially related to health, safety, and the environment that products must meet before they can be marketed in Europe.

The task of drawing up the corresponding technical specifications that meet these essential requirements of the directives is the work of the European standards bodies. Such specifications are referred to as *harmonised standards*. By following the relevant harmonised standard, they will benefit from a 'presumption of conformity' to this marking. The alternative for a manufacturer that for some reason wants to choose his/her own methods must be able to provide a so-called 'technical file' upon request.

In ISO the work on nanotechnologies takes place in ISO/TC 229 Nanotechnologies, the corresponding group in CEN is CEN/TC 352

---

[2]In addition to the ISO and CEN mentioned, there are the IEC (International Electrotechnical Commission) at the international level and CENELEC (The European Committee for Electrotechnical Standardization) at the European level. The standardisation work being performed on telecommunications also has its own organisations: the ITU-T (International Telecommunication Union) and ETSI (The European Telecommunications Standards Institute).

Nanotechnologies. These groups are dedicated to nanotechnologies, but nanotechnologies are also an important aspect of the work in some other TCs or WGs (working groups). To improve communication flow between these, liaisons are established.[3] To be effective, liaisons should operate in both directions, with suitable mutual arrangements. It is supposed to include the exchange of basic documents, like new work item proposals and working drafts (ISO/IEC, 2008:18–20).

Most standardisation works are initiated by industry, but standards can also be requested by the European Commission to implement European legislation. These are named 'mandated' standards.

CEN standardisation system costs approximately 800 million Euros per year. Eighty percent of these costs are carried by industry.

An overview of the steps for making a standard (in CEN):

(1) Someone proposes a new subject to their national standardisations body, which makes an evaluation of the subject.

(2) The national body's task includes, among other things, to check if there already is ongoing work on the specific subject.

(3) The national body also evaluates the one that proposes the subject: does she or he or it (an organisation) represent a certain breadth, and might other actors be interested; in the output and in participating in the work?

(4) Once the above things are checked, and other formal things are accepted, the national body either proposes a national standard work, or it sends a proposal for work to be done to the other standardisation bodies that are member of CEN.

(5) If five or more of the members vote in favour, a group is established.

(6) The group is expected to select a chairman at their first meeting.

(7) A standard would be expected from the group within a time

---

[3]The CEN/TC 352 is working to liaison with the following committees, groups, and technical bodies: EC DGs Enterprise Environment, DG RTD, JRC, Health and consumer products, ISO/TC 229 Nanotech, IEC/TC 113 Nanotech standardisation for electric and electronic products, CEN/TC 18, 137 — Assessment of workplace exposure to chemical and biological agents, 138 Non-destructive testing, 184, 243 Clean room tech, 248 Textile and products, 290, CEN/BT/WG 70 (CEN/STAR) CLC/SR 113 (ECT/TC 113), Standard Norge 2008.

frame of 3 years.[4] This would be formulated in a business plan, which is to be approved at a technical committee (TC) level. Usually a national mirror committee is established for interested national parties to follow the work and to meet and discuss with the nominated national experts participating in the European or international standardisation work. These meetings take place at a national level, which of course restricts the use of money for travel. This means that this form of participation would also be interesting for parties with fewer resources available.

The development of an ISO standard follows three main steps:

(1) Recognising the need for an international standard
(2) Working and negotiating the detailed specifications of the standard
(3) Getting the draft standard (Delemarle & Throne-Holst, 2010; ISO, 2010) approved formally

Like in the CEN system, in ISO also the public interest is in principle represented by delegates from NGOs (consumer organisations, environmental organisations, etc.) and national regulatory experts. Further, like in CEN, business/industry controls most of the relevant resources (time, travel money, expertise).

11.3 Review of the CEN and ISO Processes

11.3.1 Initiative and Objectives

The CEN develops standards and other documents on a large number of subjects. Formal European Norms (ENs) from CEN have a unique

---

[4]In a new regulation (1025/2012), the EU commission amends earlier directives on standardisation. In addition to addressing the need for reducing the overall development time for standards, it also stresses the need to facilitate the participation of SMEs and consumer and environmental organisations: 'Standards can have a broad impact on society, in particular on the safety and well-being of citizens, the efficiency of networks, the environment, workers' safety and working conditions, accessibility and other public policy fields. Therefore, it is necessary to ensure that the role and the input of societal stakeholders in the development of standards are strengthened, through the reinforced support of organisations representing consumers and environmental and social interests.' In addition it shall apply from 1.1.2013 (EC, 2012).

status, as they are also national standards in all its 30 member countries. By 2006, there were 12,300 Norwegian Standards, and more than 90% of these were pan-European standards (Pronorm, 2006:VII). Members of CEN are obliged to implement European Standards as national standards, that is, to substitute their national standards with the European one.

On the other hand European standards are voluntary, and as such they are not laws in Europe. However, laws and regulations may refer to European standards, and in that way they can comply with the mandatory standards (CEN, 2008b). And since they have been developed in response to market demand and through a consensus process among the interested parties that participated, there are good reasons to anticipate the widespread use of the developed standards (ISO, 2006).

At the international level, there is a consumer panel run by ISO, named COPOLCO (Committee on Consumer Policy). The committee develops recommendations for action, policy statements, guides for standards writers, and proposals for new areas of standardisation (ISO/COPOLCO, 2012). It has initiated standards on social responsibility, consumer complaints, and trade in second-hand goods (I Jachwitz, personal communication, 17th Nov. 2008).

11.3.2 Organisation

As mentioned, the yearly cost of the standardisation work in CEN is estimated at around 800 million euros. As most of these funds are from the industrial participants that take an interest in the outcome, one could be tempted to say that there is a reasonable amount of resources available for standardisation work. Other participants may view the resource situation rather differently, that is, the non-industrial participants. There are large travel costs for meetings, especially in the ISO. As an example, the second meeting of ISO/TC 229 Nanotechnologies took place in Shanghai, China.[5] In addition, there are costs for preparations and even for producing intellectual input to the process, for instance, in the form of scientific reports.

At the last meeting of CEN/TC 352, two working groups where established and convenors elected:

---

[5]The implication is not that China is far away from Norway/Europe, but that travelling costs get high because each successive meeting is heldat a new continent (Asia, Europe, Northern/South America etc,).

WG 1 'Measurement, characterization, and performance evaluation'

WG 2 'Commercial and other stakeholder aspects'

In addition, the TC 352 resolved to establish a task group on 'Nanotechnologies and Sustainability' (CEN/TC 352, 2008a).

One of the first things to be decided when establishing a standardisation group is the scope of the group. In the ISO standard of standards (ISO/IEC, 2008), it is stressed that considering the substantial financial and other resources (manpower) involved in the standardisation process, it is important that standardisation activities should '*start by identifying the needs, determining the aims of the standards to be prepared and the interests that may be affected*' (ISO/IEC, 2008:44). It also specifies that any proposed new work at least should include title, scope, purpose and justification, programme of work, the resources to be provided, relevant documents, cooperation, and liaison. This should be included in order to assess and justify the need for this specific activity.

As such, one would expect the scope and the content of the process to be rather clearly defined, and that this actually is available explicitly formulated and written down to the participants.

11.3.3 Participation

The work on developing standards in ISO is carried out by 3000 technical bodies, involving around 50,000 experts, mainly from the industrial, technical, and business sectors which have seen a need for a standard to meet actual or anticipated market demands. Also, other experts '*with relevant knowledge*' (ISO, 2006:7) may join these in the work. The experts are representatives of governmental agencies, consumer organisations, educational establishments, and testing laboratories (ISO, 2006:7).

The work in CEN on developing European standards is carried out by around 1800 committees and groups involving 400 European professional organisations and more than 60,000 national experts.

The experts involved in the work are not only nominated through the national standardisation bodies but also through the national mirror committees. In the latter, the national standardisation body would expect to be informed.

The chairman, or convenor, leads the work in the different committees or groups. The obligation to this position is stated clearly in the 'standard of standards' mentioned above (ISO/IEC 2008). This position has special responsibilities and duties: *'The chairman of a technical committee is responsible for the overall management of that technical committee, including any subcommittees and working groups. His task is to advise the technical management board on important matters relating to that technical committee, via the technical committee secretariat. For this purpose he receives reports from the chairmen of the subcommittees via the subcommittee secretariats.*

The chairman of a technical committee or subcommittee is supposed to:

   (a) act in a purely international capacity, divesting him- or herself of a national point of view; thus he/she cannot serve concurrently as the delegate of a national body in his own committee

   (b) guide the secretary of that technical committee or subcommittee in carrying out his duty

   (c) conduct meetings with a view to reaching agreement on committee drafts (see 2.5)

   (d) ensure at meetings that all points of view expressed are adequately summed up so that they are understood by all present

   (e) ensure at meetings that all decisions are clearly formulated and made available in written form by the secretary for confirmation during the meeting

   (f) take appropriate decisions at the enquiry stage (see 2.6).

   In case of unforeseen unavailability of the chairman at a meeting, a session chairman may be elected by the participants.' (ISO/IEC, 2008:13).

At the meeting of the CEN/TC 352 on 26th September 2008 in Madrid, 23 persons participated. With some exceptions, the names of the countries of the participants are given and not their organisations: Belgium (1), Czech Republic (1), Denmark (1), France (2), Germany (1), Norway (2), Spain (4), Switzerland (1) and the United Kingdom (3). The exceptions were three persons given as liaisons: one from the ISO/TC 229 Nanotechnologies, one from CEN/TC 184 Advanced Technical Ceramics, and one from CEN/TC 243 Cleanroom Technologies. Two persons came from EC-DG Joint Research Centre and one from the CEN Management Centre, which assists the technical committees in developing standard programmes (CCMC, 2012).

Interestingly, one person, A. Hermann, present at this meeting in Madrid, came from the European Environmental Citizens Organization for Standardization (ECOS). It is a consortium of more than 20 environmental NGOs established to *'enhance the voice of environmental protection in the standardisation processes'* (ECOS, 2008). The consortium focuses at 'increasing the ecological performance of products, ensuring sound measurement methods for pollutants, greening management systems in businesses and improving consumer information towards sustainable consumption' (ECOS, 2008). The chief task in achieving this aim of ECOS is to monitor the standardisation activities of the standardisation bodies both internationally and in Europe as well as the product policies in Europe. It is a non-profit association established in 2002, and it is funded by membership fees, the European Commission, the European Climate Foundation, and the European Free Trade Association (EFTA). Its secretariat is based in Brussels. The representatives from ECOS have participated in the standardisation work on nanotechnologies since 2007. In their response to a programming mandate issued by the EC, on the elaboration of a work programme to 'take into account the specific properties of nanotechnology and nanomaterials', dated January 2008, ECOS supported a standardisation programme of risk assessment and measurement of nanomaterials, especially manufactured ones. They argued that priority should be given to food, food packaging, and personal care products. The consortium sees some advantages in labelling, but is not sure if this is to be established through standardisation work.

More specifically on the standardisation concerning nanotechnology, the CEN describes some aspects in a report dated April 2008, as a response to the programming mandate mentioned above: *'the level of participation in standards activity is currently not fully representative of known stakeholders. Noticeably lacking are industry, consumer and national regulatory authority experts'* (CEN/TC 352, 2008b).[6] They see this in a perspective where more and better communication is needed on both the existence and the importance of standardisation activity. In addition they highlight

---

[6]This is now to be addressed, see footnote 4. However, ANEC, the European consumer voice in standardization, appears somewhat reluctant to what the COM (1025/2012) actually will result in (ANEC 2012).

the role standards can play in a regulatory context and the positive aspects of the role in the commercialisation of the nanotechnologies.

In this report, the authors point to a communication from the EC from March 2008. This communication deals with how standardisation in Europe can contribute to innovation in Europe. Here it is specifically mentioned that the access of all interested stakeholders to standardisation should be (further) facilitated. In particular they mention small and medium enterprises (SMEs), but also consumers *'to facilitate the uptake of innovation by the market'* (COM, 2008:8). Among their proposals we find this passage: *'the Commission encourages the ESOs to seek a fair and balanced representation of all stakeholders of standardisation'* (COM, 2008:8).

11.3.4 Result of the Processes

We referred in some length to the responsibilities and duties of the chairman, and we observed that there is an explicit expectation that agreements shall be reached, and that all points of view are summed up and understood by those present at the meeting.

The ideal of consensus is very clearly stated in all documents describing the process: Standards from the ISO, called ISO International Standards' *...are a normative document, developed according to consensus procedure, which has been submitted for vote by all national bodies and approved by 2/3 of the P-members of the responsible technical committee with not more than ¼ of all votes cast being negative'* (ISO, 2006:2).

In the earlier mentioned ISO standard of standards, four concepts that are needed for making procedures for international standards *'cost-effective and timely, as well as widely recognized and generally applied'* (ISO, 2008:6) are highlighted. Consensus are one of these (the others being modern technology and programme management, discipline, and cost-effectiveness). Regarding consensus, it is stressed that *'sufficient time'* should be taken for *'discussion, negotiations and resolution of significant technical disagreements'* (ISO, 2008:6).

In the light of this, it must be said that the rather strongly favoured ideal is one of a common voice and not one with articulated interest of certain stakeholders. Standards are to be used first and foremost by industry, and at least in Europe, they are a way of fulfilling legal requirements, as we have seen.

To determine whether ISO international standards should be confirmed, revised, or withdrawn, they are reviewed by the relevant technical committee at a maximum interval of five years (ISO, 2006:3). The same applies in CEN (2008c).

11.4 Deliberation Criteria

How can we say that standards work procedures hold up against Cohen's four criteria for ideal deliberation?

Standardisation discourse should be *a free discourse*; participants ideally regard themselves as bound solely by the results and preconditions of the deliberation process. It should be noted that some stakeholders might not want a standard work to succeed, and the filibuster tactics described below could then take place. This means that Cohen's ideal of a free discourse is not completely relevant for stakeholder deliberations in standardisation work. Participants will be expected to act in the best interest of their constituencies, mainly businesses.

Nevertheless, in spite of stakeholders' concern for their own interests, it is still mainly a *reasoned process*; parties are required to state the reasons for their proposals and respect arguments in the discussions.

Formally, participants in the standard process are *equal*. But some are more equal than others, first because of the uneven access to resources; to cover travel costs, man hours and background information and reports. Second, because of representativeness; the final weights of votes cast needed to finally approve a standard are based on the relative sizes of the population of each of the European countries, which is democratic in one sense, referring to the number of citizens being affected by the outcome. In another sense, it interferes with democratic procedures, because it makes participants unequal. So, inequality between participants is both an 'empirical' difference of resources brought into the work, and a formal difference because of voting rules and procedures.

Standardisation aims at *rational motivated consensus*. This is clearly stated in all documents, and also experienced by one of the authors when participating in such a process.

It seems as if deliberations in standards work clearly meet two of the criteria, but that they struggle a bit more with the two others. Processes have some important shortcomings with respect to the representation of certain groups, like consumers, but this is at least acknowledged within the international standardisation organisations.

11.5 Summarising Appraisal

Standardisation is a global business, and with so many people and money involved, it seems impossible to think of a world without it. It is actually hard to imagine an alternative. Those involved would insist that it would be a world with less contact, communication, and trade. Maybe a kind of mercantilist system would appear?

At the same time the system comes with a set of weaknesses and shortcomings that make it less satisfactory than one might wish (obviously dependent on your point of view). One is the dominant role of industry both in initiating the work and in financing it. Many technical committees are very much dominated by industry representatives with lots of resources at hand, whereas representatives from, for instance, consumer groups seldom would have comparable backing in both manpower or in monetary resources. On the basis of the experiences of one of the authors from the standardisation of toys in one of the working groups of CEN/TC52, there could also be room for some filibuster tactics from a strong group of participants. In this specific instance we suspected the industry of not wanting a standard in place 'too soon'.

In principle, it is important that both CEN and ISO open the participation in their work groups and committees to the general, non-business interest, whether this interest/these interests are represented by governments or by NGOs. The 'first level' of democracy is national: *'Democracy is an important principle for ISO. Every full member country of ISO has the right to take part in the development of any standard'* (Forsberg, 2010:30).

National interest, however, often tends to be of economic interest, and as such it is often identical to business interest. That is why consumer interests or environmental interests should be represented by NGOs. Here, the question of resources becomes crucial, as participation in ISO's work is not usually compensated by the national standards bodies. Costs have to be taken by the participants. *'This may effectively hinder the participation of NGOs, and achieving wider participation in standards making is therefore a continuous problem for ISO'* (Forsberg, 2010:30).

Given that both consumer organisations [Consumer International, the European Consumer's Organisation (BEUC), ANEC] and environmental organisations [Friends of the Earth (FoE), Greenpeace] have limited resources, and the number of relevant groups is limitless (3000 technical bodies in ISO, 1800 committees in CEN), the NGOs have to prioritise rather brutally if they want to make a real difference. *'In general then, if technologies are stabilized, this is because the network of relations in which they are involved — together with the various strategies that drive and give shape to the network — reach some kind of accommodation'* (Bijker & Law, 1992:10).

At least a degree of stabilisation seems to be a precondition to bring nanoscience and nanotechnology under some kind of democratic control, and here standardisation may place a central role. The partial stabilisation of nanoscience and nanotechnology, and the attempts at rational regulation of the field(s) stand out as one of the tasks where the NGO representation of the public interest is crucially important.

Standardisation work is one of the arenas where industry and consumer interests have the opportunity to meet and deliberate. Strengthening the consumer interest in relation to this work — and commitment to develop other venues of interaction and contact — will be important to provide consumers with genuine freedom of choice and thus the ability to take responsibility (Klepp & Laitala, 2010:90).

References

ANEC (2012) ANEC secretary-general speaks in Stockholm and Oslo. *ANEC Newsletter* 2012, 9 ed. Available at http://www.anec.eu/anec. asp?p=newsletter&ref=02-01

Bijker WE, Law J (1992) *Shaping Technology / Building Society: Studies in Sociotechnical Change*, Cambridge, MIT Press.CEN (2008a) 'What is a standard?'. Available at https://www.cen.eu/cen/NTS/What/Pages/ default.aspx

CEN (2008b) FAQs [online]. Available at http://www.cen.eu/CEN/Pages/ faq.aspx

CEN (2008c) 'How to get involved?' Available at https://www.cen.eu/cen/ NTS/How/Pages/default.aspx

CEN (2012) Services. Available at http://www.cen.eu/cen/Sectors/ Sectors/Services/Pages/default.aspx

CcMC (2012) The CEN Management Centre, available at http://www. cencenelec.eu/aboutus/MgtCentre/Pages/default.aspxCEN/TC 352 (2008a) *Report of the Sixth Meeting of CEN/TC 352*, held on 26 September 2008 at AENOR, Madrid, document CEN/TC 352 N 102.

CEN/TC 352 (2008b) *Commission mandate addressed to CEN, CENELC and ETSI for the elaboration of a programme of standards to take into account the specific properties of nanotechnology and nanomaterials.* Report from CEN/TC 352 Nanotechnologies. Drafted by the CEN/TC 352 Mandate Report Group, April 2008.

COM (2008) *Towards an increased contribution from standardisation to innovation in Europe. Communication from the Commission to the Council,* COM (2008) 133 Final. The European Parliament and the European Economic and Social Committee, Brussels.

Delemarle A, Throne-Holst H (2010) The role of standardisation in the shaping of a vision for nanotechnology, forthcoming special issue on nanotechnology of *International Journal of Innovation and Technology Management (IJITM). In press.*

EC (2010) M/461 – Standardisation mandate to CEN, CENELEC and ETSI for standardisation activities regarding nanotechnologies and nanomaterials. European Commission, Enterprise and Industry Directorate-General, New approach industriies, Tourism and CSR. Director Pedro Ortún. 5th February 2010. Available at http://www.cen. eu/cen/Sectors/Sectors/Nanotechnologies/Documents/M461.pdf EC (2012). Regulation (EU) No 1025/2012 of the European Parliament and of the Council of 25 October 2012 on European standardisation. *Official Journal of The European Commission* (14.11.2012). Available at http://eur-lex.europa.eu/LexUriServ/LexUriServ.do?uri=OJ:L:2012:3 16:0012:0033:EN:PDF

Forsberg E-M (2010) *The role of ISO in the governance of nanotechnology,* Work Research Institute report, August 2010, Oslo.

European Environmental Citizens Organisation for Standardisation (ECOS) (2008) Who are we, Brussels, Belgium, ECOS, available at http://www. ecostandard.org/?page_id=28

ISO (2006) *Catalogue 2006.* English version. ISO Central Secretariat.

ISO (2008) General information on ISO [online]. International Organization for Standardization. Available at www.iso.org/iso/support/faqs/faqs_general_information_on_iso.htm

ISO/IEC (2010) How are ISO standards developed? International Organization for Standardization. Available at http://www.iso.org/iso/standards_development/processes_and_procedures/how_are_standards_developed.htm

ISO/COPOLCO (2012) The consumer and standards. Available at http://www.iso.org/iso/standardsandconsumer.pdf

ISO/IEC (2008) *ISO/IEC Directives, Part 1. Procedures for the technical work*, Ninth edition. Available at http://isotc.iso.org/livelink/livelink/fetch/2000/2122/3146825/4229629/4230450/4230455/ISO_IEC_Directives%2C_Part_1_%28Procedures_for_the_technical_work%29_%282012%2C_9th_ed.%29_%28PDF_format%29.pdf?nodeid=10563026&vernum=-2

Klepp IG, Laitala K (2010) Standardisation and consumer responsibility, *Ethnologia Scandinavica, A Journal for Nordic Ethnology* 2010(40), pp. 82–93 Lund.

Pronorm (2006) *Katalog over Norsk standard og relaterte produkter fra Pronorm*, Pronorm, Lysaker, NORWAY.

Standard Norge (2008) CEN Madrid Sept. 2008. N_007 SN/K 298 Nanotechnology (National Committee).

Chapter 12

An Online Platform for Further Deliberative Processes

Francois Jegou[a] and Pål Strandbakken[b]

[a] Strategic Design Scenarios, rue Dautzenberg 36-38, BE-1050, Brussels, Belgium
[b] SIFO Consumption Research Norway, Oslo Metropolitan University, Postboks 4,
St. Olavs plass 0139 Oslo, Norway
f.jegou@gmail.com

12.1 Introduction

An important part of the task list of the Nanoplat project was to deliver and to test run a 'platform' for further deliberations, where we tried to build on the examples reviewed and analysed in this book (Chapters 3, 4, 10, and 11). It was decided that such a platform should be based on an online tool, which would make it possible to arrange deliberations on a European level, cross-nationally. Like the other deliverables, this online platform was to take account of and, if possible, to incorporate Cohen's deliberation criteria, ideally by having them built into its architecture.

The responsibility for developing the tool was left to the partner Strategic Design Scenarios (SDS) in Brussels, Belgium. The platform

Consumers and Nanotechnology: Deliberative Processes and Methodologies
Edited by Pål Strandbakken, Gerd Scholl, and Eivind Stø
Copyright © 2021 Jenny Stanford Publishing Pte. Ltd.
ISBN 978-981-4877-61-9 (Hardcover), 978-1-003-15985-8 (eBook)
www.jennystanford.com

was tested, using Nanoplat consortium members as moderators, through two stakeholder conferences on the theme 'food and nano', in May 2009. In this chapter we present the thinking and the key concepts behind the online platform. Further, we will briefly comment on the test runs, discuss some results, and indicate possible improvements.

The ambition was to develop a platform where we could arrange semi-directed online deliberations between stakeholders from production, consumption, and governance. The analysis of recent deliberative processes on nano in Europe and in the United States seemed to indicate that the role of the promoting institution is crucial. It selects and provides information to participants, tries to ensure a certain level of continuous engagement/attention among them, and takes responsibility for implementing the different steps of the process while itself staying independent of the interest groups in the field.

This platform is activated by an independent third party, the promoter. The promoter takes the initiative, defines the framework, recruits participants, and monitors the process. The Nanoplat platform requires an independent moderating institution to bring the deliberation forward. The independence and integrity of the promoting institution is obviously very important, not the least because it has to moderate the debate in real time and in the eyes of all participating stakeholders.

One intention behind the Nanoplat project was to facilitate some form of deliberative process between various players involved in the definition, production, and commercialisation of a particular class of consumer goods based on nanoscience and nanotechnology and to do it at a point in time when it might be possible to influence actors in the field. This is an argument in favour of an early 'upstream' public engagement (Delgado *et al.*, 2010). What we wanted to achieve was to facilitate a more enlarged discussion among stakeholders. The interviews conducted in different professional sectors on current deliberation practices show that this focus on a particular product value chain tends to overlap with the already ongoing regular provider–client relationships. To a certain extent, this could interfere with business-to-business strategic discussions normally protected by confidentiality. They, however, also indicate that the platform should focus on promoting deliberation between *a larger number of different types of stakeholders*, including authorities responsible for the regulation of the sector, professional associations active in the regulation of the technology (hearings, work on standards), NGOs watching consumer interests, environmental interests, etc. Deliberations could also include broader dialogues between *stakeholders of similar type in different European countries*. These may benefit from sharing experiences with themes containing potential conflicts as well as anticipating possible future regulation. The platform should be useful for both types of dialogues.

To join in a deliberation is demanding in time as well as efforts. In particular it is hard for a non-expert to acquire and maintain an updated level of necessary knowledge. To successfully maintain a minimum level of interest among participants, the promoting institution requires a continuous stimulation of the debate similar to the moderation of a round table: feeding the exchanges with renewed points of view, focusing on burning issues, and ensuring that all consistent groups of stakeholders are aware of the deliberation and have access to it. The notion of a permanent deliberation process should therefore be understood as a continuous and regularly regenerated process.

From the review of a sample of deliberative processes made by the Nanoplat consortium, it seemed reasonable to think that to develop a permanent deliberative process at a European level, it would have to be constructed as an online-based platform.[1] The use of such a platform, particularly, when it is constructed as an online social computing-like tool, should facilitate deliberation between remote participants in different European countries and support the different organisational and monitoring tasks of the institution promoting the deliberation.

In theory, such tools may enable deliberation processes where participants carry out the discussion autonomously without moderation and without third-party input. Observation of the development of social computing phenomena shows, however, that this ideal deliberation situation will occur only after the platform has existed for a certain time and has reached a certain level of development. The platform has to particularly raise interest among a critical mass of users and demonstrate that its benefits outweigh the effort needed to engage with the platform.

---

[1]Nanoplat's responsibility was not to actually construct a permanent European deliberative platform but rather to reflect on what such a platform might look like and to develop and test a model of it.

To sum up, outside the rather artificial testing situation of the project, stakeholders will only use the platform and engage in online deliberations if it is more interesting and rewarding than time consuming and more rewarding than existing alternative activities. The online deliberating platform will have to develop from an interesting experiment with stakeholder influence and participatory democracy, mainly for academics, to a useful tool for players in the field; something worthwhile to engage within itself.

12.2 Description of the Platform

Taking the deliberation criteria of Cohen (1989) as the starting point for the development of the online tool, it should facilitate *a free discourse*; participants should ideally regard themselves as bound solely by the results and preconditions of the deliberation process. Online deliberation should be *reasoned*; parties will be required to state the reasons for their proposals. The participants in the deliberative process are *equal,* and the deliberations aim at rational motivated *consensus.*

The construction of the platform should also meet a set of preconditions: It would have to enable deliberation exchanges on a European basis, and it would have to consider that it is involving high-level experts and professionals, all with rather strong time constrains. Further, the platform would have to facilitate intensive and dense interactions, meeting both the limited availability of the participants and their expectation of benefitting from the participation. The platform should aim at limiting the side tasks of organising, coordinating, and processing the information, and in order to ensure that proper deliberation could be engaged in at any moment, the process should be kept as easy as possible for participants. This means that most of these tasks should be taken care of by the promoting institution.

Taking these conditions and constraints into consideration, the actual deliberative process supported by the platform is based on two steps, *kick-off sessions* and *revision sessions.*

The purpose of the kick-off sessions was to bring out a set of key issues while involving a limited circle of experts in a quick interaction process. These sessions consisted of short online conferences, and they used a regular chat-like tool; it was keyboard-based with no audio or video, allowing short written fluid exchanges between 5 to 10 participants. One purpose of such setting was to slow down exchanges between potentially antagonist parties on controversial subjects. On the one hand, the very fact of having to type a text to participate was supposed to induce participants to a more composed attitude. Body languages and tone of voice do not appear, and moods appear only through inflexion of written formulations. Written contributions more or less require rational thinking.

These interaction settings were meant to inspire a *reasoned* debate, balancing the dynamics of a round table with the argumentation of the written paper, thus meeting the second deliberation criterion. The end result was an 8- to 12-page written dialogue that remained available online, both as an evidence of the exchanges and as a ready-to-use material to prepare a synthesis for the next step.

The purpose of the open revision session was to facilitate the emergence of an agreement within a larger circle of stakeholders. This second type of session was based on open access online revisions of syntheses that had emerged from the kick-off sessions. This process was based on a 'wiki-like' tool (explained in Section 12.3), displaying the synthesis and giving participants the possibility to edit them and to substitute the former versions by new ones. The tool also preserves and documents the history of all previous versions and the changes that have been made, with the possibility to restore them and to compare the different versions.

The revision sessions were designed like this in order to inspire consensus between different stakeholders, their points of view, and their respective interests. The synthesis from the kick-off session was already giving a balanced and reasonable position in order to meet different stakeholders' positions. For participants, the effort required to disagree is higher than to agree: the editing of an already structured text takes some time and attention, so only consistent controversy was likely to be sustained. Small divergences would just lead to fine-tuning of the text or even accepting as it was. Polemic attitudes and ego valorisation were discouraged. In this way the design of the platform met the *consensus* criterion. As we have seen in some of the other processes, consensus is a *choice*. Another option, just as democratic, could have been to encourage a tolerance of disagreement and even majority votes. For the platform, however, there were some arguments in favour of consensus. The most important is probably that it encourages the formulation of a common stakeholder position on certain issues, which is something that can be fed into politics and decision-making as a position and not just as an individual opinion.

The log of the visits allowed the moderating institution to monitor the number and type of visitors. It was possible to see if they agreed to the synthesis, if they made changes in it, or if they simply read the text and approved it .Thus, the final result was the last version of the synthesis, agreed upon by all participants.

Two other important aspects of the platform have to be mentioned to show how it met the two other deliberation criteria.

First, stakeholders were invited to take part in both kick-off sessions and open revision sessions. This invitation described the conditions of the exchanges in generic terms without even mentioning the real identity of the other stakeholders (they joined as 'business' 1, 2, 3, etc. 'research' 1, 2, 'NGO' 1, 2, and so on), leaving them free from undue influence, and meeting the *free discourse* criterion. Actors were bound solely by the logic of the deliberation.

Second, participants received a specific login code and a password, but their identity was not disclosed. Their login code only stated the category of stakeholders they belonged to. This kind of anonymity is one way of meeting the *participants' equality* criterion.

12.3   The Toolkit

The deliberative processes that were reviewed by the consortium came with various purposes, forms, lengths, sizes, etc. This showed that it would be necessary to adapt the process to specific goals and contexts. The platform was therefore organised as a set of tools that could be used in different ways, with parts in different sequence and with different intensity, in order to enable a fruitful implementation of different deliberative processes.

For this reason (flexibility), the online platform was designed as a structured toolkit that was available for actors wanting to implement customised deliberative processes. The online platform offered the following tools, all supposed to be useful for a wide range of possible deliberations:

A focused library. The platform offered a limited online library, based on a Nanoplat project theoretical discussion of the notion of deliberative processes, as well as a selection of background documents and references, accumulated by the consortium. This provided a minimum common working framework reference in terms of deliberative processes.

A catalogue of deliberative processes. A series of case studies, documented by the consortium, showed a panorama of various contemporary experiences in terms of form, duration, size, purposes, etc. In order to facilitate access, browsing, and comparison between processes in the catalogue, three different levels of description of the processes were accessible online: a diagram positioning the different processes according to the duration and number of participants involved; an intuitive visual mapping representing the major characteristics of each process in terms of inputs, profiles of the process, and outputs provided; and the complete written descriptions of the cases.

Support for the study of production-consumption-governance actors. Short phone interviews were conducted in order to gain an initial overview of current experiences with deliberative processes among various stakeholder groups. These were performed prior to more ambitious descriptions and analyses of cases. The experience gained during the Nanoplat project was made available online. The material was obtained through a series of suggestions for recruitment of candidates for interviews; good practices in the moderation of the discussion; realisation of a semi-directive interview guide; and a sample of interviews realised with stakeholders of different groups and from different countries.

Briefing documents. A logical first step was to ensure that sufficient basic information on the topic was shared among the participants as well as among promoters of the deliberation. A series of short briefing documents on nano and food, prepared by consortium members, and a selection of related reference publications were made available on the platform. Since these deliberations were expected to involve already informed stakeholders, only rather light information was provided. It should be noted that these series of short briefing documents also played a role in creating the first level of consensus between the involved stakeholders. Participants' (stakeholders') agreement on the background material should be regarded as a first step towards convergence.

'Brain food' material. SDS had prepared a very interesting set of fantasy input; beautifully designed fake advertisements for possible

future food and nanoproducts as well as fake newspaper articles on imagined technological breakthroughs. The idea is that it is helpful to open up the debate a bit, breaking with pre-formed ideas and engaging in something interesting.

Participants list. A list on a spread sheet allowed the promoting institution to follow the different levels of involvement of the invited participants as well as to match their anonymous login with the logs of the visits they made on the various parts of the platform.

A kick-off chat tool. As previously described, this tool allows a small number of participants from various stakeholders groups and different countries to exchange ideas through a written discussion and raise key topics to deliberate on. In order to achieve a fluent and interesting exchange of views, we estimated that it should ideally involve 5–10 participants in each deliberation; many enough to cover a range of interests and views, few enough to let everybody be heard.

A revision wiki tool. As explained above, the tool allowed a large number and variety of stakeholders to review and to agree or disagree with a synthetic consensus statement developed as a result of the deliberation. It is a wiki-like tool means that participants in the deliberations are invited to revise and rewrite the end document without losing the earlier versions. This was regarded as an innovative way of developing informed consensus.

12.4   Testing the Platform

The Nanoplat consortium ran two pilot experiments with the platform in May 2009. The first one, which began on 25th May, had only four 'real' users, three from NGOs and one from political authorities, plus one academic observer, a moderator, and an administrator from the consortium.[2] The second one, which began on 29th May, had six real participants; two from business, and one each from research, NGO, authority, and academica, in addition to two observers, the moderator, and the administrator. Both sessions focused on the theme of food and nano. The objective was not to be exhaustive on the topic but rather to test the platform, experiment

---

[2]Eight or ten consortium members followed both sessions, but they were not supposed to interfere.

with it, and explore its potential as well as to identify its weaknesses and point to possible improvements.

Briefing documents on the topic were developed in advance and fed into to the semi-directed online debates. They gave a common theoretical framework and synthesised the main issues in order to facilitate discussions and gave an equal initial minimum knowledge to the kick-off sessions participants.

The two kick-off sessions were organised, and four to eight production, consumption, or governance actors were invited to take part in an online deliberative process. Invitation mails outlined how the debate would be organised. We proposed a two-hour meeting, set the date of the meeting, and asked each potential participant to indicate some issues under the food and nano topic they might want to debate.

Participants had to confirm their interest areas before participating. They received an answer by e-mail stating the kind of stakeholders that would participate in the debate anonymously and their roles, or just the kind of stakeholder they were, and the kind of questions that would be debated by making clusters of 6–8 questions from all questions received. As mentioned, participants also received a personal username and password identifying their role but not their personal identity, that is, 1_business; 2_research; 3_NGO; 4_authority, etc. Participants were invited to check if they could log on the platform and familiarise with discussion tools prior to the discussion session.

Guidelines and rules of participation were shared with the participants, particularly to ask participants to systematically justify 'give reasons for' and 'properly explain' their answers, respecting Cohen's criteria of reasoned debate.

During the 2-hour meeting, the participants logged on the platform with a consortium participant as moderator. The six to eight issues proposed by the deliberation participants were debated for 10 to 20 minutes each. After the session, a one-page synthesis on each of the four initial issues was made by the consortium.

One of the (previously mentioned) deliberation mechanisms, proposed in the *Future Food Dialogue Project,* was based on a series of photo-realistic future products elaborated from the hypothesis and opportunities envisioned from the general development of nanoscience and nanotechnology in the food sector. This tentative anticipation was not an attempt to forecast the future food market but a stimulation material to foster large stakeholder deliberation on both likelihood and desirability of this conjecture. It consists of a mix of serious and naïve and reasonable and provocative hypotheses circulating in the public debate. The purpose of the deliberative debate is to draw tentative lines between realistic futures and fuzzy dreams and to share views on expected progresses and speculative or misleading wonder worlds.

The syntheses developed for the four emerging issues were posted on the Nanoplat platform via a wiki-based tool. Invitations were sent by mail to the kick-off session participants, to the observers of the session, to interested stakeholders that were not available for the kick-off session, and in general, to a larger range of production-consumption-governance actors of the focused topic. In total, 60 invitations were sent proposing to check each of the four issues and eventually to revise the related synthesis. Invitations also explained that the synthesis will be made public to incentive participation.

The consortium monitors the revision process, prompting participation and avoiding too radical interventions. Forty-six persons had been invited to participate, and 15 had effectively logged on the website over a 1-week period. The level of feedback (33%) was particularly high, considering the very short time available between the invitations and the sessions to get time from high-level experts and professionals, allowing us to consider that the synthesis was validated (Cohen's criteria of aiming at consensus).

12.5 Summarising Appraisal

The content of the debates on the topic of nano and food and the design of the interaction, seem to confirm that the Nanoplat online platform — when activated by a competent promoter — provides sufficient support to facilitate meaningful deliberations between a wide range of stakeholders from different backgrounds and across different countries. In addition, the level of participation of the experts obtained, with regard to the relative small effort to engage them in the experimentation provided by the consortium, also tends to confirm that a permanent deliberation process on nanoscience and nanotechnology at a European level can be conducted by one or more independent institutions on a permanent basis and at a relatively low cost.

The pilot experiments were too short for us to draw any in-depth conclusions on the platform, however. More piloting on a larger sample of stakeholders and different topics should be run to confirm the first results. However, it is clear already with this experiment that an online deliberation platform is a promising tool for promoting a regular dialogue between various players on the European nanoscience and nanotechnology scene and on more general technology development in Europe and beyond.

We have also observed the weak points of the tool, which should be addressed to develop the tool further:

The current beta version of the online tool would benefit from being further developed into a more stable, robust, and user-friendly infrastructure. Further, a dissemination mechanism, to give visibility to the ongoing debate, should be integrated in the platform. However, a process of translation of the experts' conjectures into potential tangible offers on the market has been further developed by the consortium to engage with large public.

A large number of online initiatives, from surveys of new products to consumer information and public forums are already running on the Internet. Building synergies with them is the key to disseminate debates supported by the platform. As a potential spin-off of the Nanoplat prototype of the deliberative platform and pilot experimentation of nano and food, a follow-up project called *Future Food Dialogue Project* has been developed in collaboration with the Responsible Nano Forum in the United Kingdom.

Whatever happens with the Nanoplat online platform in the future, we are confident that we have demonstrated how an online tool for stakeholder deliberations can be designed; it solved some problems, met some potential challenges, and pointed to some interesting problems. Like the U.S. National Citizens' Technology Forum (Chapter 8), this platform also demonstrates possible future uses of online solutions for deliberative processes.

References

This chapter builds on the day-to-day experiences with the development of the platform in SDS. In addition we use a number of internal project reports and more elaborate deliverables from work package 6 of the Nanoplat project.

Cohen J (1989) Deliberation and democratic legitimacy, in *The Good Polity* (Hamlin A, Pettit P, eds), Oxford, Basil Blackwell.

Delgado, Wickson, Kjølberg (2010) Public engagement coming of age: From theory to practice in STS encounters with nanotechnology, *Public understanding of science*, 20(3).

METHODS AND APPROACHES FOR STAKEHOLDER AND CITIZEN INVOLVEMENT

Introduction to Part IV

Pål Strandbakken
SIFO Consumption Research Norway, Oslo Metropolitan University,
Postboks 4, St. Olavs plass 0139 Oslo, Norway
pals@oslomet.no

The fourth part starts with the last chapter of the first edition, *Conclusions: Towards a Third Generation of Deliberative Processes* (Chapter 13). In this chapter, the editors envisioned a next step, a new, 'third generation' of deliberations, after observing the development of the field from the Citizens' Nano Conference in Denmark in 2004 (Chapter 3). The French Citizen Conference in 2006–2007 (Chapter 7) was defined as the 'model' second generation deliberation, with its resources, its more clearly defined aims and its rather close linkage to political authorities. In the chapter, the editors analysed how the design could be improved on and defined a set of characteristics for a future possible third generation.

Chapter 14, then, is a report from a part of a European project where SIFO had the opportunity to develop the concept further theoretically and to test it empirically. The test of the third generation concept was part of the NanoDiode project. As a theoretical refinement and an empirical test of the third generation concept it sort of 'fulfils the promise' of the original's concluding chapter.

In the very large European project Strenght2Food, SIFO led a Work Package that tested out Michel Callon's concept of 'hybrid forums' in the development of local food resources (Chapter 15). Amilien and her colleagues took some elements from the third generation deliberations and introduced them in their own design of hybrid forums, called HF 2.0. Even if it was not used for studying (or interfering with) emerging technologies like nano, but for food systems, the design appears to be transferable to science and technology. Compared to the third generation deliberations, HF 2.0 is more 'anarchic', less predictable because it is much about having heterogeneous groups defining and developing a common project. It aims more at social change than at producing statements.

The ultimate chapter (Chapter 16), called 'Conclusion 2020: A More Democratic Science Through Public Engagement?' is an assessment of the status of deliberations after the 'turn to RRI' post 2010. Further, it compares the third generation deliberations with Strenght2Food's version of hybrid forums, called HF 2.0. Finally, it takes a new look at the relation between deliberative, or participatory democracy and the representative system, building on the experience from the analysed cases.

Chapter 13

Conclusions: Towards a Third Generation of Deliberative Processes

Eivind Stø,[a] Gerd Scholl,[b] and Pål Strandbakken[c]

[a] *National Institute for Consumer Research (SIFO), Postboks 4682 Nydalen, 0405 Oslo, Norway*
[b] *Institute for Ecological Economy Research (IÖW), Potsdamer Str. 105, 10785 Berlin, Germany*
[c] *SIFO Consumption Research Norway, Oslo Metropolitan University, Postboks 4, St. Olavs plass 0139 Oslo, Norway*
eivind.sto@sifo.no

13.1 Introduction

Throughout this book, various processes where citizens and/ or stakeholders have been invited to deliberate on the future of nanoscience and nanotechnology have been described and analysed. The aim of this chapter is to draw some conclusions after eight years of experiments with deliberative processes on these themes and to discuss the content and design for a future generation of deliberation over emerging technologies in general and nanoscience and nanotechnology in particular.

First, we will discuss the development from what we have chosen to call the first to the second *generation* of deliberative processes.

Consumers and Nanotechnology: Deliberative Processes and Methodologies
Edited by Pål Strandbakken, Gerd Scholl, and Eivind Stø
Copyright © 2021 Jenny Stanford Publishing Pte. Ltd.
ISBN 978-981-4877-61-9 (Hardcover), 978-1-003-15985-8 (eBook)
www.jennystanford.com

This discussion is based upon the theoretical framework and overview of cases presented in Part I of this book as well as on the in-depth studies of selected processed reviewed in Parts II and III.

These conclusions drawn will constitute the starting point discussion about the potential future of deliberative processes in this field. Will more exercises in participatory democracy only produce more of the same reflections and just repeat a limited set of political concerns and suggestions, or do we need a *third generation*?

Finally we suggest some areas suitable for public involvement, areas we believe to be rather soon on the political and scientific agenda.

13.2   The Generational Perspective

We introduce the notion of 'generations' of deliberative processes as a means of describing a kind of learning process across time and space, and as a way of conceptualising a development towards increased sophistication up to the present and into the future. In this perspective, if some ways of approaching the themes of nano deliberation seem less relevant today it is not because of any process design faults, but because they probably no longer produce new and interesting insights.

At this point, we will just present a very broad overview of how we distinguish between generation one and two. One should bear in mind that the differences between generations are about main tendencies and not clear-cut definitions. Obviously, the generational perspective is applied after the events. The organisers of the Citizens' Conference in Île-de-France never knew that they would later be seen as pioneers of a new generation of deliberations.

Deliberations placed under generation one are necessarily rather general, often focusing on the role of science in modern societies, how it is financed, and how it should be controlled. Their take on nano is also rather general, since they were conducted at a time when there were much talk about unlimited potentials, unspecific possible risks, and few real applications. The image of nano was closely linked to the science laboratory. In addition, organisers at the time would have to assume that the average citizen knows nothing about 'nano' and perhaps has never even heard the expression. Also, the relationship between the deliberation and world of political decision-making was unclear.

So, the second generation is defined as a gradual development from these constraints. They would address more specific areas, like medicine/health, food, or cosmetics. The gradual appearance of at least some consumer products, and the debate over more specific near-future applications tended to move focus from the laboratory and over to society. Second generation deliberations tend to be given a more explicit political role, where organisers have some ideas about what results and contributions from the processes should be used for, even if the direct application of suggestions and demands often appear to be difficult.

13.2.1 The First Generation of Deliberative Processes on Nanotechnology

The first generation of deliberative processes are often characterised by addressing general ethical, political, and social aspects of nanotechnology and by focussing on the role of science in modern society. The processes are concerned with transparency in the financing of nanotechnology research. Furthermore, the processes are addressing general questions; they are not very specific neither as far as technology nor applications is concerned. With some exceptions, it seems reasonable to date the period of this first generation from 2004 to 2006.

The Citizens' Nano Conference in Denmark (Chapter 3) and the U.K. NanoJury (Chapter 4) definitely belong to the first generation of deliberative processes. This is also the case for the NanoTechnology Citizens' Conference in Madison, USA (Chapter 8). However, it is also possible to identify a large number of other relevant processes, as we have seen in Chapter 2.

As early as 2004, the Australian Commonwealth Scientific and Industrial Research Organisation (CSIRO) held a one-day workshop with community members, nanotechnology specialists, CSIRO staff, and government representatives to explore citizens' views on the social, economic, and environmental implications of nanotechnologies (the Bendigo Workshop on Nanotechnologies). Discussions in working groups were stimulated by scenario kits and revealed a mix of optimism and concern among the participants with respect to nanotechnologies. Their benefits were particularly appreciated in the context of enhancing socioeconomic well-being and environmental sustainability (Mee *et al.*, 2004). CSIRO used the findings to draft a 'community issues checklist' which helps researchers and research planners reflect on the social, economic, and environmental issues linked to nanotechnology from the citizens' perspective.

As a follow up to the Bendigo community engagement workshop, CSIRO organised a Citizens' Panel in Melbourne in December 2004 (Katz *et al.*, 2005). The participants were citizens from the local community and civil society organisations. In the morning sessions, the presentations of six invited speakers were discussed by the lay panel. In the afternoon, three break-out groups, taking the roles of community, industry, and government, discussed the issues further and formulated group positions as a response to the question 'What statement will Australia make to the United Nations Forum on Nanotechnology in 2006?' Participants were, among other things, concerned about ownership and control of emerging technologies, the adequacy of regulation for nanomaterials, and the social divides that nanotechnology might generate. They were in favour of any nanoapplication contributing to the decoupling of resource consumption and economic growth, and they stressed the need for democratic accountability and transparency in science and technology research and development. CSIRO used the findings of the Citizens' Panel and the Bendigo workshop to develop recommendations for nanotechnology research and future social research on nanotechnology.

As seen, the Danish Citizens' Conference in 2004 belongs to the first generation of deliberative processes on nanotechnology. The initiative to 'Citizens attitudes to nanotechnology' was taken by the Danish Technology Board (DTB), who also financed the process. This was the first deliberative process in Denmark on nanotechnology. However, DTB had carried out similar processes earlier, focusing on other topics, and Denmark has a strong tradition for attempts at participatory democracy. This method of citizens' involvement was developed from 1987, but since then various changes and improvements have been introduced.

The time line of this event was short. The meeting took place on the evening of 7th June 2004 in the three hours from 17:00 to 20:00. Participants only met this evening. Prior to the meeting they had received some written material on nanotechnology, limited to 13 pages of popular information. The topic of this meeting was general and not limited to specific areas or technologies. The discussions concentrated on visions for nanotechnology on the one hand and on ethical, social, and environmental aspects on the other. No specific consumer goods were discussed, but concern over consumer products were a part of the discussions.

The aim of this process was the 'representation of a common voice'. The goal of the process was not to reach to any form of consensus. This was a *citizen's* conference, not a *consensus* conference. The participants were not selected because they represented specific stakeholder interests. It is important to bear in mind that this conference took place in 2004, which is reflected both in the questions put forward and in the answers from participants. We are not talking about specific advice to public authorities but more about input to the general political discourse.

The results from the conference were reported to the Danish Ministry for Science and Technology. The ministry used the experiences and the report from this process very actively in their strategic plan for research on nanotechnology and nanoscience in Denmark in December 2004. The plan explicitly discusses ethical and environmental risks with reference to citizens' involvement. These reflections later became a part of the future research program in Denmark.

The following year, NanoJury UK was established (Gavelin *et al.*, 2007). Initiated by Cambridge University Nanoscience Centre; Greenpeace U.K.; the Guardian newspaper; and the Policy, Ethics and Life Sciences Research Centre (PEALS) of Newcastle University, this public engagement exercise aimed to influence policymaking by systematically building and articulating public opinion on nanotechnology. Twenty-five randomly selected citizens formed the jury, although the jury was accompanied by a multi-stakeholder oversight panel, which monitored balance and fairness of the process, and a scientific advisory panel ensuring proper presentation of evidence. The oversight panel recruited the experts ('witnesses') who informed the jury on the relevant matters.

During the first half of the process, eight evening sessions of two-and-a-half hours each, the jury explored a set of general, non-nano social issues: young people, social exclusion, and crime in the local community. In the second half, consisting of 10 sessions of two-

and-a-half hours each, the jury focussed on nanotechnologies. The finishing sessions were dedicated to writing recommendations on the future development of nanotechnologies in the United Kingdom. These recommendations, each presented with figures of support by the jury, were presented to an audience of policymakers, researchers, and journalists.

Amongst other issues, the jury called for more openness on public spending on nanotechnology research, for publicly funded research to focus on solving long-term environmental and health problems, and for all nano-enabled products to be tested for safety and proper labelling. Commenting on the outcomes of the NanoJury, Doubleday and Welland (2007) concluded that the exercise fed into a wider process of policy learning from public dialogue on nanotechnologies, exemplified by the collection of evidence, by the British Nanotechnology Engagement Group (NEG), from engagement projects such as the NanoJury, and the subsequent reporting of these to the UK government (Gavelin *et al.*, 2007). In addition the process made the involved scientists more conscious of the wider social and political contexts of nanotechnology research.

The Madison Area Citizens' Conference on Nanotechnology, held in April 2005, represents the first major public engagement exercise on nanotechnology in the United States (Gavelin *et al.*, 2007; Kleinman & Powell, 2005). It was organised by the University of Wisconsin's Center on Nanoscale Science and Engineering and its Integrated Liberal Studies Program. The process took place over three Sunday meetings and involved a group of 13 citizens from a variety of backgrounds. All participants received background information before the first meeting, which was dedicated to the preparation of a list of questions about nanotechnologies. The second meeting was held as a public forum, where seven experts from a range of different fields responded to the questions of the citizens' panel. The final meeting was devoted to drafting recommendations for government, which were later presented to the public at a press conference. Amongst other things, the recommendations related to health and safety regulations (e.g., testing of nanomaterials), media coverage and information availability (e.g., databases and product labelling), research and research funding (e.g., increased funding of research into social and ethical implications), and public involvement (e.g., effective mechanisms for citizens' involvement in nanotechnology policy development). Whether any concrete action has been taken on the recommendations remains unclear (Gavelin *et al.*, 2007:123).

13.2.2 The Second Generation of Deliberative Processes on Nanotechnology

The differences between the first and second generation are not dramatic. We have, however, seen a gradual shift and we focus on two interesting changes. First, the processes tend to get more specific. They no longer just address general aspects of nanotechnology but deal with specific areas like cosmetics or medicine. Second, we see a change of focus from the laboratory to the society. This second generation is also, to a degree, concerned with challenges for the consumer market.

Among our in-depth cases, the two French processes seem to build a bridge between the second and the first generation. The aim of the Citizens' Conference in Île-de-France (Chapter 7) was to influence the research policy of the region, and at the same time, its contents were more specific than those of its predecessors. This also seems to be the case for the German Consumer Conference (Chapter 5) and for the U.S. National Citizens' Technology Forum on Human Enhancement (Chapter 9). These second generation events were arranged from around 2006.

For the Île-de-France process, the explicit purpose was to generate advice to improve the quality of the decision-making process related to nanoscience and nanotechnology research in the region. The initiative was taken by the political authorities who also financed this rather expensive and time-consuming process. The participants were selected as citizens, and their task was to represent the common voice. The process lasted for 4 months and included three training weekends.

There are two reasons to consider this process in the second generation. First, the participants addressed more specific themes. They were given an active role in the development of the process and used the possibility to put specific topics on the agenda and to invite experts to present them. They ended up with the following topics: medicine, the economy, the environment, communication, and defence. The second reason is more problematic, and may take this actual process beyond the second generation towards a possible third generation; its close link to political process in Île-de-France. We will return to this below.

One year after the Madison citizens' conference, the German Federal Institute for Risk Assessment (BfR) conducted a 'Consumer Conference on the perception of nanotechnology in the areas of foodstuffs, cosmetics and textiles' (Chapter 5) as part of its risk communication activities. The consensus conference involved a consumer vote on recommendations of how to deal with nanotechnologies in the selected domains (Zimmer *et al.*, 2007, 2008). A group of 16 citizens was introduced to the subject through background material disseminated prior to the first meeting and at two preparatory weekends of lectures and discussions. On the basis of these inputs, the lay panel was asked to prepare a catalogue of questions on consumer-related aspects of applications of nanotechnologies in foodstuffs, cosmetics, and textiles. In parallel, the group chose experts for a public hearing from various stakeholder groups (science, public agencies, industry). After this hearing, the group prepared its vote in private deliberation. The next day, the vote was presented to the public and handed over to representatives of the government and civil society organisations.

The vote calls for comprehensible labelling, clear definitions, terms and standards for nanomaterials, and for more research into potential risks before nanotechnology is used to a larger extent in consumer products. The vote names foodstuffs as the most sensitive area for the use of nanomaterials. Regarding the use of nanotechnology in cosmetics and textiles, however, the consumers felt that the foreseeable benefits clearly outweighed potential risks. The BfR took a number of initiatives to disseminate the consumer vote among decision makers. They presented the vote at scientific conferences and to the German 'Nano-Kommission' (a multi-stakeholder board), the consumer committee of the German Bundestag, federal and regional authorities, industrial associations, and to the European Food Safety Authority (EFSA).

An evaluation carried out after the process arrived at the conclusion that the Consumer Conference was, by and large, a transparent process of deliberation (Zimmer *et al.*, 2008). Consumers and experts had a clear understanding of their roles, and the entire process was transparent to outsiders through extensive press coverage. It is less clear, however, what, if any, impact the consumer vote had on decision-making in policy, science, and business. The initiators regarded it as a pilot in public engagement and have not conducted a similar exercise since.

The UK Nanodialogues, a project led by the think-tank Demos and the University of Lancaster, consisted of four experiments with upstream public engagement, run throughout 2006 (Gavelin *et al.*, 2007; Stilgoe, 2007). The first experiment, a 'People's Inquiry on Nanotechnology and the Environment', comprised three workshops with a group of 13 residents of East London, and concentrated on the use of nanoparticles to clean up chemically contaminated land. The second, 'Engaging Research Councils', involved citizens, scientists, and research council staff and aimed to explore and discuss the role of public engagement in research planning. The third, 'Nanotechnology and Development', was run as a three-day workshop in Zimbabwe. It examined how nanotechnology might help local communities to secure clean water. The fourth experiment, 'Corporate Upstream Engagement', was based on a series of four consumer focus groups discussing nanotechnologies in hair products, oral care, and food. It was run in cooperation with Unilever and tried to explore the potential of public engagement for corporate research and development.

For this concluding chapter the second experiment, Engaging Research Councils, is most relevant (Chilvers, 2006; Stilgoe & Kearnes, 2007). It involved three sessions; two full-day meetings and a final workshop on the preparation of conclusions and recommendations. The process started as two groups (one with six full-time mothers with children of school age and the other with eight young professionals with an interest in technology), which were merged into one for the second session. In the first session participants were made familiar with nanoscience and nanotechnology, and the role of research councils and were asked to prepare questions for discussion with scientists and experts during the second session to be conducted two weeks later. The final session suffered from poor participation of lay participants from the previous workshops (4 out of 14). Due to the experiment's focus on early-stage research the recommendations addressed broader issues of science, technology, and society. They advocated clear and easy-to-understand language in public science dialogues, the involvement of the public at all levels of the research process, and an intensification of public engagement on nanotechnologies. The evaluation of this second experiment concludes that the process did not meet initial expectations of encouraging public engagement and the delivery of final recommendations:

> Rather than its potential to shape future directions in nanotechnology research per se, it seems that the real value of this experiment lies in its possible influence on learning and reflection within the Research Councils (and other scientific institutions) about the role of public engagement in shaping research agendas in nanotechnology (and other areas of science) (Chilvers, 2006:11).

In Switzerland, a major public deliberation process on nanotechnologies was the publifocus discussion forum on Nanotechnology, Health, and Environment (Rey, 2006). It was organised by TA-SWISS (the Swiss Centre for Technology Assessment), a publicly funded body for the assessment of emerging technologies with a record of using participatory methods, and funded by the Federal Office of Public Health (FOPH), the Federal Office of the Environment (FOEN), and the Zurich University of Applied Science Winterthur (ZHW). The discussion forum aimed to explore how citizens perceive nanotechnologies in the context of health and environment. The publifocus consisted of four focus group discussions with citizens (53 in total), carried out in different regions in Switzerland (Winterthur, Bern, Lausanne, and Lugano) in September 2006.[1] The groups were recruited to represent gender as well as different occupations, educational levels, and social and political interests. Each group discussion was four hours long, starting with introductory talks from two scientists covering technical and societal perspectives on nanoscience and nanotechnology. After these presentations the participants discussed the topic in two one-hour discussion blocks. Right after the event the participants filled out feedback forms, which were used for evaluation purposes. The main outcome of the process was a report by TA-SWISS (Rey, 2006), the findings of which were fed into nanotechnology policymaking, particularly at FOPH and FOEN.

The U.S. National Citizens' Technology Forum, conducted in the United States in March 2008 and funded by the U.S. National Science Foundation, was a deliberative process run simultaneously across

---

[1]Additionally, a fifth focus group was carried out with stakeholders from various national organisations and associations.

six different sites in the United States — New Hampshire, Georgia, Wisconsin, Colorado, Arizona, and California. It was initiated by the Centre for Nanotechnology in Society at Arizona State University (CNS ASU) and coordinated by collaborating partners at North Carolina State University (Hamlett *et al.*, 2008). The process aimed to generate informed, deliberative public opinion on how to manage technologies for human enhancement in order to demonstrate that non-experts can come to informed judgements on complex issues if they have access to adequate information and to provide a good example of public engagement that may help ordinary citizens to voice their interests and contribute to shaping public policy.

At each of these sites, panels of lay citizens — roughly representative of local demographics — were recruited to discuss, debate, and give recommendations on converging technologies for human enhancement (i.e., nanotechnology, biotechnology, information technologies, and cognitive science; NBIC). Since these technologies have not yet delivered a wide range of commercial applications, the process addressed an early stage of technology development. The Citizens' Technology Forum involved a total of 74 citizens completing questionnaires about their knowledge and views on these technologies before and after the process, reading prepared background material, discussing and debating what they saw as the important issues, formulating and asking questions of invited experts in the field, and developing a final report with recommendations for policymakers on how to manage these new technologies. There were face-to-face meetings within the individual groups on the first and last weekends of the month while interactions across the different groups occurred in nine two-hour online sessions held throughout the month. Researchers from a university at each location served as coordinators and facilitators for the individual groups. The lay citizens received $500 upon completion of the process.

13.3 Conclusions on Deliberative Processes on Nanotechnology

The main insights of these reviews of exercises in public engagement in the domain of nanotechnologies can be summarised as follows:

There is a *wide spectrum of organisations driving public engagement on nanotechnologies*, including academia (universities, research institutions), policy consultants, policy-advising research bodies, professional engagement facilitators, public authorities, and research councils. Different initiators use these processes to impact decision-making to different degrees — from informing the general public and/or stakeholders to funding research — which, of course, influences the potential *impacts* of the deliberative process.

There are also different *purposes* behind deliberative processes. They can be about the general identification and assessment of public attitudes towards a certain technology, about experimenting with a new form of public dialogue in order to learn about its potentials and shortcomings, or about informing a specific decision, for example, on research funding, from citizens' perspectives. In some cases the idea of *experimentation* with novel forms of public engagement has been important: thus the question of how the process can be organised in an appropriate fashion comes into focus. This reveals that public participation and deliberative processes do not follow a given format. Rather, different *forms* of deliberative processes are used, from 2-hour card games on nanotechnologies to single evening events, focus group discussions of 3-hour length, and processes running over 6 months with three weekends as face-to-face contact and additional interaction in between these meetings. Accordingly, there is a *variety of tools employed to stimulate interaction* between participants, such as working groups, public hearings, plenary discussions, presentation plus question and answer session, scenario techniques, and card games.

The *results* of the deliberative processes reviewed are numerous. There are *direct and tangible* ones, encompassing votes, recommendations, and reports. *Indirect and intangible* ones include participant learning, such as awareness of and sensitivity to the opportunities and risks of nanotechnologies, learning how to manage and employ deliberative processes, and building trust into public risk assessment and management. The *actual impact(s)* of the depicted deliberations, however, are difficult to assess due to a lack of data, specified goals, and information about dissemination activities. If *policymakers* are only loosely linked — or not linked at all — to the deliberative process, the actual impact on (their) decision-making is obviously very small. This appears to be the case in deliberative processes driven by academia (such as the U.S. National Citizens' Technology Forum). Thus a *prerequisite* for a significant impact would be the description of a clear avenue for how the deliberative process is going to influence policymaking; often, one encounters no such description.

13.4   The Future of Deliberative Processes

In the discussion about the future of deliberative processes in the case of nanotechnology we have seen increasing doubts about further development of deliberative approaches. It is possible to identify arguments along three dimensions:

- New processes will probably not create more knowledge but will more or less reveal more of the same.
- The increased use of deliberative processes will raise public expectations, but these expectations will not be met by occasional processes in which no one has a more permanent responsibility.
- An increased use of deliberative processes will be a threat to numerical and representative democracy. The processes move the power of decisions from governmental institutions to non-representative processes not designed to make such political decisions.

We will argue for a third generation of deliberation in nanotechnology. In order to do so we have to meet the arguments stated above, and to discuss these three dimensions of scepticism. *More of the same?*

In our overview of selected deliberative processes we saw a movement from first to second generations of deliberation. Besides a chronological difference, the distinction between first and second generation deliberative processes on nanotechnologies is most evident in terms of the sophistication of the methodology used.

Second generation exercises are more elaborate than early approaches. To some extent they are also more specific processes: they deal with particular applications rather than the general relationship between science and society. Responsibility is thus moved from the research community to industry. This also has to be reflected by the third generation of deliberative processes on nanotechnology. As we see it, the next generation of deliberative processes need to be even more specific. The reason for this is the fact that we are no longer talking about nanotechnology, but about nanotechnologies. This means that it is a bit less meaningful, from the ethical, legal, and social aspects (ELSA) perspective to carry out general processes.

Secondly, during the last few years a large number of nanoproducts have reached the consumer market. More than 1000 products are listed in the Woodrow Wilson Center's updated inventory, ranging from sports equipments to textiles and from cosmetics to car polish products. This is also an argument for more specific processes where strategic areas of the consumer market can become a topic. In the 2004 citizens' nano conference in Denmark, participants were uninterested in nano consequences for the consumer market: this would probably not be the case in 2009. One important emerging area is nanofood, and another related area is nanotechnology for food packaging. However, as innovation takes place, other product categories may soon be relevant.

13.4.1 Unfulfilled Expectations?

In terms of impact, however, it is difficult to draw a clear-cut distinction between the two generations. While there is a lack of knowledge on this topic, the link to political decision-making appears to remain fairly weak.

One of the challenges for deliberative processes is that they create substantial expectations amongst citizens — particularly for those who participate in them. What will happen with our input? Who is responsible for the voice of the public in the future? This is a serious concern, because some of the processes are parts of research projects, and the deliberation ends with the project. Others are parts of public programmes, which also close at the end of the programme. As an example, the Danish Technology Board carried out deliberative processes or stakeholder involvement in 2004, 2006, and 2008, but involvement of citizens was not on the agenda in other years.

Future deliberative processes have to deal with these challenges. In the Nanoplat project, we tried to deal with this challenge by establishing a web-based platform for deliberation (Chapter 12). This has a potentially more permanent character, and it may be used in future processes.

Engagement in social computing processes is facilitated when participants find forms of reward or gratification. The first level of kick-off sessions assumes the form of a round table, allowing participants to debate with peers across Europe and to benefit from the discussion. The second level of open revision sessions gives access to an up-to-date level of consensus between stakeholders.

These two elements are already promising benefits for fostering engagement between participants. To add to these incentives for engaging in the deliberation platform, the Nanoplat consortium proposes adding visual forms of representation (showing scenarios that may result from the deliberative process) to the different levels of written synthesis. The scenarios developed for the platform propose a visual synthesis through the design of some hypothetical products in line with agreements reached by the deliberative process. They are intended to express a balanced position, somewhat challenging compared to the current situation, but reasonable and justified.

The purpose of these visualisations is then twofold: on the one hand, it should stimulate contributors to the deliberative process by showing them a concrete expression of the consequences for the future resulting from what they have agreed. On the other hand, it should facilitate access to the debate by a larger number of stakeholders by translating the debate into the form of concrete — if still hypothetical — products.

13.4.2 A Threat to Numerical Democracy?

We are aware that there may be some perceived tensions within deliberative processes in general, and in the development of deliberative processes on nanotechnology more specifically. These processes represent increased citizen involvement in democratic processes. Both in the United States and in Europe we have, over the last decades, seen deliberative and stakeholder approaches to the governance of GMOs and nanotechnology. The Danish Board of

Technology developed a model for public involvement in complicated technological processes by the 1980s, and this model has helped create legitimacy for other deliberative processes.

But there are also renewed discussions about the relationship between representative democratic processes and these participatory deliberative processes. In public and academic discourse we have seen an increasing scepticism to some aspects of the deliberative processes. Who participates in them, and what is the goal of the processes? Within political science this has relevance for classical discussion of numerical democracy and corporate pluralism (Rokkan, 1969).

It is possible to identify at least three key challenges for an inclusive, democratic debating, and decision-making process on new technology: (1) knowledge deficits amongst participants and stakeholders, (2) the discrepancy between hypothetical visions and actual commercial products, and (3) how and if will the outcomes of such debates be fed back into decision-making processes?

A lack of knowledge about nanotechnology is also documented in scientific research. This is surely the case for the public in general but also among political actors and other stakeholders. The consequences are that there are a limited number of voices heard in the public debate.

While the predominant representation of nanotechnology in popular science and in the media often comes as fiction relating to micro-machines and assemblers (Crichton, 2002; Drexler, 1986; Gibson, 1996), the presence of nanotechnology in ordinary life is more about carbon nanotubes in sport equipment, nanoparticles in cosmetics, and antibacterial clothing and kitchen equipment. This discrepancy between nanovisions and nanoreality makes it difficult to define a set of themes around which to organise a debate. However, experience has shown that it is possible to engage the public in relatively complicated scientific discourses.

One last challenge is the link to democratic decision-making processes. Is the deliberative process part of a set of inputs to decision-making processes so that the results are brought directly and formally back into this process? Or is it more a part of research projects in which the results are inputs into political and scientific discourses but not directly linked to formal processes? We have accounted for both kinds of processes in this book.

One critique from both participants and organisers of deliberative processes has been that it is problematic that deliberation is not a part of formal political processes. It is easy to understand this critique. On the other hand, when we are talking about numerical democracy and deliberative processes, it is also problematic when the results are brought directly back into political processes because of the diversity of the subject, lack of knowledge, and the biased representations within these processes.

13.4.3 An Answer to This Critique

The answer to this critique is that we have to distinguish between public discourse and formal decision-making processes. Deliberative processes have given a positive contribution to democratic discourse on science in general, and on nanotechnology more specifically. This represents no threat to democracy — the opposite is actually the case, because it increases public involvement and represents a democratisation of science. However, when we move to formal decision-making, we have to take all decisions within the framework of representative democracy, in which one man and one woman each have one vote.

Thus, the third generation of deliberative processes probably will have to combine public engagement with more formal stakeholder involvement. These stakeholder involvements may take various forms: *Codes of Conduct* may be one actual instrument and standardisation processes (Chapter 11) may be another.

Standardisation processes play and important part in the global economy and encourage global technological solution. Standardisation is crucial for emerging technologies such as nanotechnology. Standards are created by bringing together all interested parties such as manufacturers, consumers, and regulators of a particular material, product, process, or service. In principle standardisation could be regarded as a deliberative process with main contributions from all relevant stakeholders. However, it is not a free discourse, and the participants are definitely not equal. Industries participating in the standardisation processes will, of course, defend their own economic and technical interests. They are also more powerful in most cases than representatives from the consumer organisation. Experiences have shown that it is possible for NGOs to influence certain processes, but it is time-consuming and consumer organisations lack the necessary resources to defend the interests of consumers.

Regulation of emerging technologies like nanotechnology can take different forms. It could go from moratoriums like some governments, including Norway, imposed on the cloning of animals and humans after the appearance of the first cloned sheep Dolly. At the other end of the regulatory spectrum, there could be no specific regulation at all, like for the Internet, with a few limited exceptions though, like pornography and domain names (Hodge *et al.*, 2007). In between these two extremes we can find a varying degree of voluntary and mandatory measures, including traditional hard law approaches or self-regulation, implying forms soft law (Abbott & Snidal, 2000; Hodge *et al.*, 2007).

Since there currently are no regulations of nanotechnology, standardisation plays an important part as a soft law regulatory tool, together with various Codes of Conduct. We argue that for an emerging field, the issue of standardisation does not only rely on the production of normative documents that will be used by regulators. The notion of vision shaping is also central. To develop this point, we use the framework of a *field configuring event* (Chapter 12 and Delemarle & Throne-Holst, 2012) in the analysis of the case study on nanotechnology. We first go back to the role of standardisation and show the extent it is relevant in an emerging field context. We then present the methods and case study that we discuss in the next section before concluding.

EU has developed Codes of Conduct for responsible nano-science and nanotechnology, mainly directed towards the scientific community as guidelines for their scientific work. These Codes of Conduct have one element in common with the standardisation process. They are both voluntary soft-law tools, developed through deliberative processes between stakeholders. They represent the legal element in the ELSA program. It is a guideline developed for all stakeholders involved in nanoscience and nanotechnology research and development. However the main stakeholders are within the scientific community and involved in the technical innovations from science to applications. One relevant topic for stakeholders' involvement could be participation in the reformulation of the existing Codes of Conduct.

The scope of the Codes of Conduct is relatively broad. It covers all nanoscience and nanotechnology activities undertaken by the European research areas. Codes of Conduct invite all stakeholders to act responsibly and cooperate with each other, in line with the N&N Strategy and Action Plan of the Commission, in order to ensure that N&N research is undertaken in the Community in a safe, ethical, and effective framework, supporting sustainable economic, social, and environmental development.

This broad approach also means that the Codes of Conduct cover nanoparticles, nanosystems, nanomaterials, and nanoproducts, within the scale of 1–100 nm. It is a voluntary guideline directed towards all relevant stakeholders. However, it is our impression that the targeted groups belong to the scientific community.

The precautionary principle plays an important part in the Codes of Conduct. These should not only apply for researchers but also for other professionals, consumers, and citizens and be relevant for the environment. It is worth noting that research organisations also should apply good practices as far as classification and labelling is concerned.

The Codes of Conduct are also concerned with the dissemination strategies and activities related to the principles and guidelines formulated in the document. The success of this voluntary guideline also depends upon knowledge about these principles among large groups of individual researchers and approval by research institutions, Member states should support the wide dissemination of these Codes of Conduct, notably through national and regional public research funding bodies. In addition to the existence of these Codes of Conduct, N&N research funding bodies should make sure that N&N researchers are aware of all relevant legislation as well as ethical and social frameworks

To sum up our argument: there is a future for a third generation of deliberative processes in the development of nanotechnology. These processes need to be more specifically oriented and more closely linked to decision-making processes. They may gain from using the platform developed within the Nanoplat project. One of the main challenges in the future is the question of who will take the responsibility for running such processes and independent institutions may be one answer to this question. Deliberative processes represent a democratisation of science, and, as long as we distinguish between public discourse and formal decision-making processes, deliberation represents no threat to numerical democracy.

The main elements in the third generation of deliberative processes on nanotechnology are specific processes where there will be a fruitful combination of classical bottom-up deliberative processes with stakeholder involvement through hearing systems, standardisations, and Codes of Conduct. This is how these processes may respond to the critique that they represent a threat to numerical democracy, and how the links to the formal decision-making processes are taken care of.

13.5 Relevant Topics for Deliberation

We have been arguing for more specific processes. We will very briefly discuss three actual processes:

- Nanomedicine
- Nanotechnology in food and food packaging
- Labelling of nanoproducts in the consumer market

Nanomedicine: A hot topic for deliberation

There are all reasons to believe that ethical, legal, and social aspects with nanomedicine will soon enter the political agenda, and that this topic is suitable for citizens' involvement through various form of deliberative processes.

The reason for this is obvious. It is possible to identify clear visions for the positive role of nanomedicine in both the diagnoses and treatment of various severe diseases, including cancer, diabetes, Parkinson's and Alzheimer's diseases, and cardiovascular problems. We are talking about a potential scientific and technological breakthrough of substantial importance. This breakthrough could improve the quality of life in an ageing European population. Robert Freitas believes that nanorobots may improve diagnoses and inspections of the human body and make it possible to repair, reconstruct, and improve element of our cells (Freitas, 2005). It could be possible to search and destroy the very first cancer cell before it develops into a tumour.

At the same time, we have to face difficult ethical challenges dealing with the use of resources on the practical/political level and the potential to add a new dimension to the human body on the other. Even the most positive contribution may have unintended consequences. We may improve the human body beyond any known quality for many of the human functions, where increased life expectancy could be one of the dramatic changes. We will comment briefly on the ethical challenges identified by the European Union (EU) commission (2006):

- *Non-instrumentalisation.* In medical research, are human beings used as a means rather than as an end of their own? We have also seen protests against using animals in tests and experiments.
- *Informed consent.* This is closely linked to the principle above. The freedom to choose and to know about realistic alternatives. At least patients need to know if they are part of a research program with control groups.
- *Non-discrimination.* Do we all have the level of treatment, or is it possible to identity differences between social groups? This may be a crucial question in the early phases of emerging technologies.
- *Equity.* This is not only a matter of national equity, but also about global equity. Will differences between the rich and poor worlds increase? This may have dramatic consequences for the quality of life and for life expectancy around the globe.
- *Privacy.* The right to privacy may be threatened by new means of diagnoses, and the possibility to combine registers. For many diseases, the correct treatment depends upon the combining register but technology opens substantial new possibilities.
- *The precautionary principle.* All these principles could be a theme for a specific deliberative process. If one should take our recommendations seriously, one has to choose among the topic listed above. However, there are also arguments supporting a broader approach to nanomedicine: there may be conflicts between the topics, for example, between personal freedom and economic resources. Let us illustrate these dilemmas.

We have seen that consumers are more sceptical to cosmetics than to sport equipment based on nanotechnology (Throne-Holst &

Stø, 2008; Stø & Throne-Holst, 2006). They are also more sceptical to nanofood than to car wax and more sceptical to underwear than to outdoor jackets. It seems like they are protecting their body. Here, the precautionary principle seems very important. When it comes to medicine, especially to experimental cancer the risk and potential benefits. This is the conclusion from several Norwegian focus groups. People are willing to take large risks when they are seriously ill.

In addition to these principles, there is an interesting ethical discussion about human enhancement, often referred to as 'playing God' (Peters, 2008). It is not only a matter of repairing the human body but improving the functions. Peters define enhancements as 'manipulating the intricacies of human nature so that human nature becomes something other than it is. We mean changing nature'.

This definition could be an excellent point of departure for a deliberative process among citizens about bio- and nanomedicine. Human beings have always been changing their nature, and there are examples of improving the human body that we accept, when applied/used under certain conditions. In the report 'Ethics of Human Enhancement' prepared for the U.S. National Science Foundation, the authors (Allhoff *et al.*, 2009) use one illustrative example. We accept to use steroids to help muscular dystrophy patients to regain their strength, as a part of a therapy. But it is strictly forbidden to use the same means to improve your strength as an athletic. We will also accept Ritalin to improve concentration for ADHD patients but are opposed to using the same medicine for ordinary students.

The fundamental question is how to draw a borderline between ethical enhancement of human bodies that we may accept and enhancement that are not acceptable. There are reasons to believe that this borderline is not fixed, but is flexible and changing all the time. It is possible to identify differences between cultures, generations, social classes, and gender?

We accept vaccination programs and other medical innovations to improve the quality of life and enhance life expectancy for human beings. Thus, we have witnessed substantial improvements during the last century. However, will we accept a life expectancy of 200 years? Of which reasons will we reject such visions?

We accept improving mental performance to a certain degree by using training programs and pharmaceutical that is available today. But to what degree will we applaud an integration of information and communications technology (ICT) into the brain? There are reasons to believe that computer chips in our brains will cause discussion among scientists, policymakers, and among lay people.

Cosmetic surgery is one of the increasing medicine businesses in modern societies. To some degree it is also socially accepted to improve your attractiveness by such means. However, this technology may soon improve the function of the human body substantially: artificial eyes, ears, and noses that surpass the capability of our natural body. Are we playing God? These are questions that have to be discusses not only among scientists and political authorities but also among individuals in their role as citizens, patients, and consumers.

Nanotechnology in food and food packaging: A salient topic for deliberation

Studies have shown that consumers are sceptical to nanotechnology in food products, compared with medicine, surface treatment, electronics and clothes. The results of a German study show that people are most sceptical to military application and food (Grobe *et al.*, 2008). Another study shows that 75% of the respondent in a German study would buy nanotextiles but only 20% gave the same answer for food (Zimmer *et al.*, 2008). In a Norwegian study we found that consumers had no negative feelings with nanotechnology for sport equipment, car care products, and electronics. However, as we approached the skin, the scepticism increased for clothes and especially cosmetics (Throne-Holst & Stø, 2008). With the GMO food discussion in mind, ELSA related questions are highly relevant for both food products and food packaging.

There is a growing use of nanotechnology in food packaging. An estimate of this market puts the total market value at $360 m for 2008. This is projected to develop into $20 bn by 2020 (Cava & Smith, 2008).

Currently, nanomaterials in food packaging are used for reducing gas permeability, adding anti-bacterial properties, blocking UV radiation, for informing to the consumer on the quality of food, and for slip surface technology.

The rapid increase in the number of nanomaterial applications in food packaging industry has raised many question about safety, ethical, policy, regulation, and environmental issues (Helland & Kastenholz, 2008). Nanotechnology still has a big challenge to create novel material in future. The research and development in this area has to take into account all additional safety consideration besides reaching all packaging requirements.

Labelling of nanoproducts: The right to information and the right to choose

As we have stressed in this chapter, consumer products based on nanotechnology are reaching the consumer market in increasing extent. More than 1000 products have been identified as nanoproducts in the business–consumer dialogue, and there are all reasons to believe that the number is much larger in the business–business communication. In addition, many consumer products are probably not advertised as nanoproducts, even though they are based on nanotechnology.

This is one of the reasons for the scientific and political discussion about labels. During the last years we have witnessed an interesting discussion about the labelling of consumer products made by nanotechnology. This is a strong candidate for a deliberative process where individuals are mobilised in their role as consumers.

The backgrounds for these demands have been that many consumer products, based on nanotechnology, lack relevant information and labels. This missing information is problematic because it challenges two of the main consumer rights: *The right to information and the right to choose* (Kennedy, 1962). The right to be informed is specified as the right to be given the facts needed to make an informed choice and to be protected against dishonest or misleading labelling. In many situations information is a necessary precondition for choice. Relevant for this discussion is also the possibility to choose alternatives to nanoproducts.

Labelling of consumer nanoproducts is problematic, because we are talking about a large spectre of difference technologies. This has been an argument against nanolabels during the last years. However, within the food sector and for cosmetics the demand for labelling schemes has increased in strength in the recent discourses. This claim has been supported by important actors such as Friends of the Earth, The European Consumer's Organisation BEUC, and Which?, the consumer association in the UK. Furthermore, it has also been supported by the Transatlantic Consumer Dialogues (Falkner *et al.*, 2009).

However, are nanolabels on cosmetics a solution or do they create new problems? Labels are a part of the toolkit for regulating the global market for cosmetics. The right to information is one of the fundamental consumer rights, and it is difficult to see any legitimate arguments against such labelling regimes. It is supported by research and also confirmed in our focus groups. It is also one of the demands from joint global actions from consumer and environmental NGOs.

However, one argument against labels may be that very few consumers have heard about nanotechnology and that they will not understand the label. An even stronger argument is that large consumer groups even will misunderstand the label. However, labels may also function as an informative tool that will educate consumers and increased knowledge will make is it easier to choose.

In EU cosmetics are regulated by the 1976 Cosmetics Directive. This is under revision, now as a EU regulation. A draft proposal from the EU Commission was in March 2009 discussed in the EU Parliament. One of the articles calls for the EU Commission to publish a catalogue of all nanomaterials used in cosmetics and establish mandatory labelling schemes for nanomaterials in cosmetics. This new regulation will probably enter into force in 2012, and this will change the regulatory regime for nanocosmetics in Europe.

On the other side, too, to which degree will this regulation create new problems? It will solve some of the problems related to consumer trust, but we can see two potential problems with this new labelling regime. The first is related to the definition of nanomaterials and the second to the relationship between the regional and global regulations of nanoproducts.

The definition of nanoscale is limited to 100 nanometres (nm), or below. This means that materials in the scale of 100–500 nm is not a part of the new regime and will not be labelled. The risk assessment within nanotechnology cannot today conclude that 100 nm is the crucial size, and that regulation of larger scale is not necessary. Thus, BEUC has called for a redefinition of the nanosize to open for potential challenges in the 100–300 nm size.

The second problem is related to the global dialogue. It is difficult to regulate nanotechnologies on the national, and even on the regional European level. We have seen in the transatlantic dialogue between the United States and EU that the labelling schemes proposed by the EU will not gain support from the U.S. government. EU has also failed to export their Codes of Conduct for research in development of nanotechnologies. And it is not a surprise that parallel positions were taken by EU and the United States in the GMO discourse. Thus, this may contribute to a new WTO conflict between the two main actors in the world market. Furthermore, if the Asian actors do not agree on potential labelling regimes, we may have to wait for a possible standardisation within ISO. This will certainly take time.

13.6   Summarising Appraisal

If we believe that a more democratic science is desirable and possible, the development of more mature or sophisticated forms of emerging technology deliberations is an interesting approach. They can contribute to a heightened awareness of questions of science and technology in society, produce input to science policy, and be used for articulating citizens' concerns over perceived risky technologies back to decision makers and manufacturers.

Participatory democratic techniques are, however, no substitute for the regular democratic processes. The ideas phase will use other methods than the then decision phase. But if we want to consider, respect, and include the voice of the common man and the voices of different stakeholders, deliberative processes do deserve a place in the policymaking of modern societies.

And with a sustained focus on upstream engagement, aiming to influence processes early on, we might have to use citizen and/or stakeholder deliberations in order to simulate a public debate that is not yet there.

For such purposes, we believe that thorough processes, respecting the deliberation criteria and taking on very specific questions, with clear and explicit linkages to a decision-making level will be an excellent instrument.

References

Abbott KW, Snidal D (2000) Hard and soft law in international governance, *International Organization,* 54(3), 421–456.

Allhoff F, Lin P, Steinberg J (2009) Ethics of human enhancement: An executive summary, *Science and Engineering Ethics,* Springer Link, available at   http://files.allhoff.org/research/Human_Enhancement_ ES.pdf

Cava D, Smith A (2008) One small step, available athttp://www.packaging-gateway.com/features/feature43794/

Chilvers J (2006) Engaging research councils? An evaluation of a Nanodialogues experiment in upstream public engagement, Birmingham, University of Birmingham, available at http://www.bbsrc.ac.uk/web/FILES/Workshops/nanodialogues_evaluation.pdf

Crichton M (2002) *Prey*, London, Harper Collins.

Delemarle A, Throne-Holst H (2012) The role of standardisation in the shaping of a vision for nanotechnology, *Special issue on nanotechnology of International Journal of Innovation and Technology Management* (IJITM). Accepted for publication.

Doubleday R, Welland M (2007) NanoJury UK, Reflections from the perspective of the IRC in Nanotechnology & FRONTIERS. n.p., available at http://www.nanojury.org.uk/pdfs/irc_frontiers.pdf

Drexler EK (1986) *Engines of Creation: The coming era of nanotechnology*, New York, Anchor Books.

Falkner R, Breggin L, Jaspers N, Pendergrass J, Porter R (2009) Consumer labelling of nanomaterials in the EU and US: Convergence or divergence?, EERG briefing paper 2009/03, London, Chatham House.

Freitas RA (2005) Current status of nano medicine and medical nano robotics, *Journal of Computational and Theoretical Nanoscience*, 2, 1–25.

Gavelin K, Wilson R, Doubleday R (2007) Democratic technologies? The final report of the Nanotechnology Engagement Group (NEG), London, available at http://www.involve.org.uk/assets/Publications/Democratic-Technologies.pdf

Gibson William (1996) *Idoru*, Penguin Books.

Grobe A, Renn O, Jaeger A (2008) *Risk Governance of Nanotechnology Applications in Food and Cosmetics.* Report prepared for IRGC.

Hamlett P, Cobb MD, Guston D (2008) National Citizens' Technology Forum Report, CNS-ASU Report #R08-0002, n.p., available at http://cns.asu.edu/files/report_NCTF-Summary-Report-final-format.pdf

Helland A, Kastenholz H (2008) Development of nanotechnology in light of sustainability, *Journal of Cleaner Production*, 16(8,9), 885–888.

Hodge GA, Bowman DM, Maynard AD (2007) *International Handbook on Regulating Nanotechnologies*, Edward Elgar Cheltenham, UK; Northampton, MA, USA.

Katz E, Lovel R, Mee W, Solomon F (2005) Citizens' panel on nanotechnology. Report to participants, DMR-2673, CSIRO Minerals, Clayton South, Australia, available at http://www.minerals.csiro.au/sd/pubs/Citizens_Panel_Report_to_Participants_April_2005_final_110.pdf

Kennedy, President John F (1962) Special Message on Protecting the Consumer Interest, March 15. Washington DC: Government Printing Office.

Kleinman D, Powell M (2005) Report on the Madison area citizen consensus conference on nanotechnology, n.p., available at http://www.nanocafes.org/files/consensus_conference_report.pdf

Mee W, Lovel R, Solomon F, Kearns A, Cameron F, Turney T (2004) Nanotechnology: The Bendigo Workshop, DMR-2561, CSIRO Minerals, Clayton South, Australia, available at http://www.minerals.csiro.au/sd/pubs/Public%20report.pdf

Peters T (2008) *The Stem Cell Debate.* Fortress Press.

Rey L (2006) Public reactions to nanotechnology in Switzerland. Report on publifocus discussion forum 'Nanotechnology, Health and the Environment', Bern, available at http://www.ta-swiss.ch/a/nano_pfna/2006_TAP8_Nanotechnologien_e.pdf

Rokkan S (1969) Numerical democracy and corporate pluralism, in *Political Opposition in Western Democracies* (Robert A. Dahl), 70–115.

Stilgoe J (2007) Nanodialogues: Experiments in public engagement with science, London, available at http://www.demos.co.uk/files/Nanodialogues%20-%20%20web.pdf

Stilgoe J, Kearnes M (2007) Engaging research councils: Report of an experiment in upstream public engagement, (Draft 3, March 2007), London, available at http://www.epsrc.ac.uk/CMSWeb/Downloads/Other/NanodialogueEngagingResearchCouncilsReport.pdf

Stø E, Throne-Holst H (2006) The precautionary principle in nanotechnology – Who should be precautionary? The role of stakeholders in the governance of nanotechnology, Paper presented at the ESA-workshop Sociology of consumption, Durham, UK, 1 September 2006.

Throne-Holst H, Stø E (2008) Who should be precautionary? Governance of nanotechnology in the risk society, *Technology analysis & Strategic management*, 20(1), 99–112.

Zimmer R, Domasch S, Scholl G, Zschiesche M, Petschow U, Hertel R, Böl G (2007) Nanotechnologien im öffentlichen Diskurs. Verbraucherkonferenz zur Nanotechnologie, *Technikfolgenabschätzung. Theorie und Praxis*, 3/2007, 16. Jg., 98–101.

Zimmer R, Hertel R, Böl G-B, (Hrsg.) (2008) BfR-Verbraucherkonferenz Nanotechnologie. Modellprojekt zur Erfassung der Risikowahrnehmung bei Verbrauchern, Berlin, available at http://www.bfr.bund.de/cm/238/bfr_verbraucherkonferenz_nanotechnologie.pdf

Chapter 14

Third Generation Deliberative Processes on Nanotechnology

Pål Strandbakken

SIFO Consumption Research Norway, Oslo Metropolitan University,
Postboks 4, St. Olavs plass 0139 Oslo, Norway
pals@oslomet.no

14.1 Introduction

In Chapter 13, we proposed that a next development step should be a 'third generation' of deliberative processes as a tool for making science more democratic. Such future deliberations were supposed to 'contribute to a heightened awareness of questions of science and technology in society, produce input to science policy, and be used for articulating citizens' concerns over perceived risky technologies back to decision-makers and manufacturers'. In the NanoDiode

Consumers and Nanotechnology: Deliberative Processes and Methodologies
Edited by Pål Strandbakken, Gerd Scholl, and Eivind Stø
Copyright © 2021 Jenny Stanford Publishing Pte. Ltd.
ISBN 978-981-4877-61-9 (Hardcover), 978-1-003-15985-8 (eBook)
www.jennystanford.com

project,[1] SIFO had the opportunity to develop the third generation concept theoretically, in addition to testing it empirically. SIFO, as a front runner, arranged a 'model' deliberation, and the institute was the 'task leader' for five further deliberations arranged by consortium members from Germany, the Netherlands, Austria, Italy, and France.[2]

In this chapter, we first describe some of the theoretical considerations prior to the NanoDiode deliberations. Then we describe the Norwegian event on human enhancement in some detail. Further, we comment on the events hosted by the European consortium members, before finally summarizing the experience and comment on the status of deliberative processes today. The chapter ends with a list of minimum requirements for a third generation deliberative process on nanotechnology, that is, 3GDP (third generation deliberation).

14.2 Theoretical Status of Third Generation Deliberations Prior to the Tests

The NanoDiode project aimed at establishing an innovative, coordinated programme for outreach and dialogue throughout Europe in order to support effective governance of nanotechnologies. The project tried to integrate engagement initiatives along the whole value chain, combining 'upstream' public engagement with 'midstream' activities and 'downstream' strategies. The overall objectives were the following:

> To develop new strategies for outreach and dialogue along the value chain, to organize engagement and dialogue at the 'upstream' level of research policy, to enable processes of co-creation during research

---

[1]*NanoDiode — Developing Innovative Outreach and Dialogue on responsible nanotechnologies in EU civil society* was a coordination and support action under the 7th Framework Programme. The project was running from 2013 to 2016 and included 14 partners/beneficiaries from the Netherlands (3), France (3), Belgium (2), Germany, Italy, UK, Austria, Poland, and Norway. The project was led by IVAM UVA in the Netherlands and it focused on the stakeholder engagement and dialogue, attempting to contribute to the responsible development of nanotechnologies in Europe.

[2]The chapter mainly builds on NanoDiode's Deliverable 3.1 — Report on third generation deliverable processes (www.nanodiode.eu).

and innovation. Further, to professionalize nanotechnology education and training and to establish an innovative program for outreach and communication on nanotechnologies. Finally, to assess the impact of the project's activities and provide policy feedback with a view to Horizon 2020.

The NanoDiode project was built on a series of previous European projects on responsible governance of nanotechnologies. The different activities and strategies were organized under four headings: INSPIRE – engagement and dialogue at the policy level, CREATE – co-creation during R&D, EDUCATE – professionalizing education and training, and ENGAGE – outreach and communication.

The project's Work Package 3, as part of CREATE, was focusing on stakeholder involvement in research and innovation processes. It sums up what we should call NanoDiode's 'midstream' activities or engagement. To the extent that downstream governance is mainly about acceptance of developed technologies, upstream governance primarily is about engaging stakeholders (and citizens) at the policymaking level, and midstream engagement is about involving stakeholders (including citizens as stakeholders) in the choices made at the level of research and development — what specific technologies, products, and applications to proceed with under the overall perspective of democratizing science and Responsible Research and Innovation (RRI). IVAM led Work Package 3 that consisted of three tasks. Tasks 3.2 and 3.3 were led by IVAM; Task 3.2 was about 'Establishing "user committees" for specific, large scale, near-application research projects in nanotechnology' and Task 3.3 was about 'Enabling a process of "regulatory research" for effective risk assessment', while Task 3.1 was about 'Developing and carrying out "third generation deliberative processes" on nanotechnology' and was led by SIFO (then the National Institute for Consumer Research, Oslo, Norway[3]).

Previously, as the consortium leader of the 7th FP NANOPLAT project, SIFO and a set of partners had reviewed projects and initiatives aiming at involving citizens and stakeholders in participatory exercises or 'deliberative processes' on the emerging nanotechnologies between 2004 and 2010. A general observation

---

[3]Now as part of Oslo Metropolitan University, the institute's official name is *SIFO (Consumption Research Norway)*.

regarding the field of deliberation was that through the years there had been a development of these initiatives from the rather modest Danish Citizens' Nano Conference 2004 — a one-day event organized by the Danish Board of Technology with 29 citizens from the Copenhagen area (Chapter 3). The Danish conference, building on years of experience with the so-called 'consensus conferences', which had been a frequently used tool in the development of that country's democratic culture, could be compared to one French event 2 years later — the Citizens' Conference, Île-de-France, arranged from 2006 to 2007 (Chapter 7). This was a regional initiative with fewer participants, but with incomparably larger time resources, money, and networks. We had observed that an increase in resources over time was a tendency.

It seemed as if, in addition to increased resources, something changed in the specificity of the tasks and in how the events were situated politically. There were also differences in the intensity of the deliberations, varying with the number of participants and the length of the process (Chapter 2).

Empirically, we identified two 'generations' of deliberative processes on nanotechnology. The first generation was mainly addressing general ethical, political, and social aspects of nanotechnology and focusing on the role of science in modern society. Processes were much concerned with transparency in the financing of nanotechnology research. The output from these events was not very specific. At the time, nanotechnology was mainly presented as a field of unlimited, but rather unclear future possibilities and risks, and few real applications.

Such general exercises were probably necessary at that time and in that technological landscape. As mentioned, when reflecting on similarities and differences between these processes, we identified a change over time in the scope and ambitions of these events. This development was described as an evolution from a first to a second generation of deliberative processes. However, 'one should bear in mind that the differences between generations are about main tendencies and not clear-cut definitions. Obviously, the generational perspective is applied after the events' (Chapter 13).

The early deliberative processes on nanotechnology were characterised by a rather broad outlook on technological possibilities and an even broader outlook on potential societal aspects. From the (participatory) democratic point of view, arrangers basically seemed to have intended the workshops to represent some sort of common voice. The resources employed were rather modest. The Danish event lasted for three hours, and the 29 participants had received 13 pages of written material in advance.

Over time, we observed that questions tended to become more specific, not any longer addressing just general aspects of nanotechnology, but dealing with specific areas, such as food, energy, cosmetics, or medicine (Chapter 5). Further, there seemed to be a slight change of focus from the laboratory to the society, even to consumer markets.

We defined the Île-de-France process from 2006 to 2007 as the 'model' second generation deliberation because we observed some novel features: The participants were expected to provide input to the science policy of the Île-de-France region on topics they were defining themselves. The process lasted for four months, with three training weekends plus a public debate and a final session to prepare a set of policy recommendations. In addition, the 16 participants did a lot of reading homework (Chapter 7). To sum it up, the characteristics of a second generation deliberation, as opposed to a first, were mainly three:

1. It had more specific themes/questions (even if it still was rather general).
2. There were more resources employed, such as time and access to expertise.
3. It had a clearer link to political processes (as something more specific than providing input to 'perhaps someone might read it' — reports to the parliament).

We took the analysis one step further, however, and asked what kinds of problems and limitations had *remained* even after these developments into a second generation. What challenges were we able to identify with the (then) state-of-the-art public nano deliberations and what constraints should be overcome if we wanted them to be even more effective tools for participatory democracy in the techno-scientific field?

- First, empirically, there was a problem connected to a replication of results and the fear that more processes would only reveal 'more of the same'.

- Second, one feared that deliberations would raise public expectations, or the expectations of the participants, expectations that were not likely to be met by the political bodies. Some of the interesting outcomes from the Île-de-France workshops could not be acted on, because the suggestions went beyond the responsibility and authority of the Regional Council and should have been addressed at a national level.

- Third, there was a democratic problem if decision-making in effect was moved from elected bodies to non-representative ones.

Those were the challenges that we thought a third generation of deliberations should have to deal with. The question of the relation to representative democracy was a theme for Chapter 1. So was, to an extent, the problem of rising and unmet expectations.[4] So, in this chapter, we focus mainly on the problems of replication of responses after a superficial addressing of the following two others:

1. *Rising expectations.* Citizens and stakeholders should enter the process with a realistic assessment of its potential effects on policy. This means that the political framing of the deliberation should be clear from the start. Is the process aiming at delivering input to the political processes, is it aiming at increasing the reflexivity of key stakeholders, or is it aiming at something else? We do not want to see participants frustrated by a lack of action in the aftermath of an interesting event that promised to have wider ranging effects than envisaged.

2. *The relation to representative democracy.* Even more important is the question of democratic legitimacy. There is always the danger that small exercises in participatory democracy might be used by authorities to bypass representative democracy. The proper democratically legitimate use of citizen or stakeholder deliberations should, however, be to produce input, ideas, and

---

[4]They are also discussed in Chapter 13.

suggestions for the representative political system to decide and legislate on. It is interesting and valuable to simulate societal debates in the form of deliberations, where such debates fail to appear. But legitimate law making, regulation, and decision-making should remain in the representative democratic system.

14.2.1 Adjusting the Design for NanoDiode Purposes

In the layout of the NanoDiode project, there were some ambiguities or uncertainties about *who* to involve in Task 3.1: stakeholders or citizens. The DOW text mentioned 'lay people in their role as citizens and consumers, or professional users of nano-enabled products' (NanoDiode Document of Work, internal project note, p. 12). It also mentioned groups that are difficult to reach, low-income groups, and so on. As it seemed inconvenient to mix the hard to reach groups, like immigrant women, with professional stakeholders, and there were not enough resources to do a double set of deliberations, we realized that we had to choose between lay people or citizens and stakeholders. The project consortium decided on testing the third generation methodology on *stakeholders*, as this was supposed to be more in line with the overall design of the project. This transfer of the approach from citizens to stakeholders did not appear to be a problem; on the contrary, it seemed interesting and potentially fruitful.

Further, there was some disagreement over the choice of themes for the deliberations. SIFO's original idea was that a common theme should be taken by all six task participants: Germany, Austria, the Netherlands, Italy, and France in addition to Norway. This design was changed, however, by the consortium early in the project. The main reason for this was that it was perceived as impossible to arrive at a theme that would be relevant and interesting in all the countries. The suggested theme — nanomedicine and human enhancement — was considered irrelevant and 'old hat' for some of the countries. So, it was decided that each task partner was free to select a theme that was considered interesting in their own specific national context. This decision might have increased the chances for each deliberation to be successful, but it made it more challenging to compare the events.

In the design, we envisioned a model with two workshops featuring (if possible) the same participants in both, being arranged a couple of weeks apart. However, the consortium decided that partners were free to decide whether they wanted to do a one- or two-day event. In the end, Germany (The University of Stuttgart) and Italy (AIRI) arranged one-day deliberations, while Austria (BioNanoNet) arranged a two-day workshop on consecutive days. Norway and France had approximately two weeks between events.

As it turned out, only the Norwegian deliberation qualified as an elaborate 3GDP; the other events were arranged with different degrees of deviation from the model. Most of them yielded some interesting results, however, and so we might claim that they ended up testing different aspects of the original design.

14.2.2   SIFO as Front Runner

For a smooth introduction to the approach and the method, it was decided that SIFO should arrange a deliberation and report from it before the other task partners held theirs. Thus, we hoped to reduce any unclear aspects of the 'third generation' concept and to test the model in a way that made it possible to adjust things before it was implemented at a larger scale. In addition, we were able to provide a template for the reporting and subsequently for the comparison of deliberations.

For the NanoDiode series of deliberations, we wanted the design — and the analysis — to focus on two distinct, but connected, aspects; *novelty of results* and *workshop democracy*. In our vision of a third generation deliberation, we had the ambition to achieve a certain degree of novel insights/statements/results, we wanted to have a high level of reflexivity concerning potential and real political impact, and we wanted to design processes where participants took responsibility for defining workshop content and approaches. In order to make the processes more relevant for stakeholders and citizens, hopefully increasing their 'ownership' of the event and the themes, we insisted on introducing a high level of workshop democracy. We know from experience that deliberations on nano and other emerging technologies run the risk of coming up with rather unoriginal and predictable recommendations or statements. Hence, we wanted all task participants to comment on this problem and outline how they aimed to counteract it. This does not necessarily mean that they should regard all replication as a failure. Participants should be allowed to shape their own perspective and to shape it stepwise. The 'standard responses' usually come rather early in the process and could be seen as a necessary first step. However, we still wanted arrangers to look for novel perspectives and ideas. It should be possible to take one step more after having arrived at the common perspective.

Further, for the third generation deliberative processes, we wanted to have participants influencing the flow of events and the choice of (sub)themes and to choose the types of expertise needed. In this way, we hoped to increase participants' feeling of ownership of the processes. In addition, workshop democracy was a means for making the deliberations more relevant and interesting for the participants.

Still, there is a need for addressing political impact. Impact can be achieved in different ways. We might have an assignment from national or local political bodies, we might influence important stakeholders, and we might initiate media debates and so on. In an earlier European project, we had struggled with a concept of political impact that we felt was too narrowly defined or operationalized because 'policymakers' was defined as persons with a defined role (elected) in the representative democratic system, such as members of parliament, representatives at EU level, representatives in municipalities, or governmental representatives. In the perspectives of 'governance', however, it seemed necessary to regard stakeholders as potential political actors.

14.3 National Report, Norway

On the 2nd and 16th of June 2014, SIFO arranged a 3GDP in the shape of two workshops in Oslo as part of the NanoDiode project, Work Package 3, Task 3.1. As mentioned, the Norwegian deliberation was supposed to be a kind of 'front runner', a model for a series of replications to be performed in the Netherlands, Germany, France, Austria, and Italy.

Stakeholder engagement and dialogue is essential to a responsible development of nanotechnology in Europe. The theoretical approach is described earlier; here, we just recap the basic elements: the three main characteristics or aims/ambitions with identifying and arranging a 3GDP on nanotechnology:

1. We wanted to go from general ethical questions into more specific political, social, or legislative and regulatory questions on more specific applications or subfields.
2. We wanted to address the problems with unmet expectations of participants by directly or indirectly connecting to political or social processes.
3. We aimed at designing a process that, if possible, could avoid a plain replication of outcomes.

The degree of specificity (1) is clearly higher than in most of the deliberations analysed here (Chapters 3–9). As for unmet expectations and policy impact (2), we have reflected on and analysed this dimension. First, we have a degree of policy relevance, formally, because the project is funded by the Commission. Further, when we define policymakers a bit wider than representatives of elected bodies and political parties, we see a real potential for connecting to political processes through influencing the involved NGOs, organizations, research administrators, and academics. Such actors will, at different points in time, be asked to respond to, reflect on, and make their voice heard in political process, be invited to participate in hearings, and so on. As such, the stakeholder-based 3GDP may have an indirect, but potentially important connection to political processes. As for the problem of replication of outcomes, we highlighted this prior to and during the workshops.

SIFO decided to do the deliberation on the themes of *nano-medicine and human enhancement* (HE). This was based on an expectation that these themes would be of interest morally, politically, socially, and economically, in addition to their fascinating technical-medical aspects. Nanomedicine is high on research agendas throughout Europe and globally. There are hopes, hypes, and expectations for breakthroughs. Some expect that the issue of nanomedicine is so value laden that deliberative processes would have given very foreseeable results — extremely positive because of the unlimited promise plus due to willingness of patients confronted with potentially lethal illness to try new therapies. But one might anticipate that using nanomedicine for 'improving' the healthy would be more controversial. This made the subject enticing in itself.

We used nanomedicine as an example of the potential of HE. This is a contested theme as it resonates to something deep inside us.

In the national context, we rather unexpectedly saw our theme appearing on the agenda of a set of civil society organizations (CSOs) and some national health institutions at the time of the workshops. This was a coincidence, but it made it easier to introduce the underlying theme or dilemma of medicine, either we will use our limited resources for curing the ill, or we will be serving and modifying the healthy (and wealthy?). Three media reports appeared in the national press in the days before the conference:

(i) One ethical dilemma discussed in Norwegian media immediately prior to our workshops was the (non-nano) regulation of the height of girls, focusing on questions of inconvenience and on normality. Who decides that a height of 185+ centimeters is not 'normal' for girls, when that is the height they will get if their growth is left uninterrupted? How do we assess the degree of inconvenience for girls who are that tall?

(ii) June is the month for exams at Norwegian universities. A student newspaper received significant national attention after a survey and a subsequent set of interviews with students using Ritalin to increase their memory. To what degree will this put the 'clean' students at a disadvantage? This is also non-nano by the way.

(iii) In the daily news round-up of the Norwegian Broadcasting Corporation, there was a report on 3D printing of organs, unlike the others, this theme has nano relevance.

This meant that thematically the deliberation got a flying start; it could refer to both examples of non-nano enhancement that were debated already and the coming wonders of nanomedicine. It was rather easy to see that nano-based enhancement might be much more dramatic than the ones used to introduce the theme.

Practicalities

The workshops were arranged at SIFOs offices in Oslo, Norway. With some minor exchanges of representatives, approximately 10 stakeholders attended. The event had representatives from the Norwegian Research Council, the Norwegian Church, the Norwegian

Humanist Organization, three representatives of the User Committee of the health region (different patients' groups), the School of Theology at the University of Oslo, the Norwegian School of Sports Medicine and the Faculty of Dentistry at the University of Oslo, and one representative from the firm Oslo Medtech.

To recruit (enough) interesting stakeholders and to have them appear and participate in this rather demanding event is a hard job and a real challenge. Interesting people tend to be busy. If they are contacted too early, there is a chance that they forget the invitation, and if they are contacted too late, their schedules tend to be filled up. The arranger can also count on last minute cancellations for one of the workshops or even both. In order to have a robust deliberation, one will need to overbook. One might need 15 confirmed participants in order to have at least 10 on both days, and at least eight persons that participate in both workshops.

This is a common challenge for stakeholder conferences, and as such it is nothing new. It can be prevented or reduced by having an interesting and important set of themes or perhaps some sort of celebrity attendance: If a well-known lecturer or a powerful politician or business leader is persuaded to attend, it will obviously ease the recruitment work.

Material

A week prior to the first workshop, SIFO distributed a 10-page booklet, introducing the themes and the approach in advance. First, the theme of Human Enhancement was introduced through non-nanomedicine, non-nano mainly to show that the moral dilemmas were here and now and did not depend on some future science fiction technology. As mentioned, the use of Ritalin for improving the cognitive performance of healthy persons without a diagnosis was one and the regulation of the height of girls was another. Second, it had to introduce the nano theme, both as a popular science presentation of the infinitely small and as an outlook on future health medicine possibilities. Third, the material had to explain what a deliberate process was and what would be the expected contributions of the participants.

The brochure or booklet was in Norwegian, as was the main parts of the workshop. For the so-called poster session, SIFO had produced two posters with content to be discussed on day one.

The posters, also in Norwegian, were on the themes Deep Brain Stimulation and 3D Bio-printing of human organs. In addition, the more general NanoDiode project poster was displayed.

Name tags with the name and relevant affiliation of each participant were handed out to the participants on arrival. We also had a PowerPoint Presentation from Harald Throne-Holst, introducing the theme and the NanoDiode project, and some written questions that were distributed before the group sessions. All material was printed in the NanoDiode template for graphic presentations.

Programme/session outlines

The deliberation was arranged as two half-day (12:00–18:00) meetings — the second coming two weeks after the first. It started with the registration, introduction to the theme, and introductions of participants and their organizations' approach to the theme. There was a poster session, where participants gathered around posters to discuss their themes. As mentioned, one poster targeted Deep Brain Stimulation and the other focused on 3D Bio-printing. Participants spent 30 minutes per poster and did the poster session as a plenary. There was a video lecture chosen by the participants and a planned 'walkshop', to continue debates outside, but this had to be skipped because some group work took too long. Day one ended with a plenary presentation and discussion of the results from breakout groups. One representative from each group presented the outcomes of the group discussions. A central point was what the participants wanted to highlight for the second day. Did they want arrangers to hire other kinds of experts; did they want to focus on other themes or approaches?

The second day started with registration and 'a quick round across the table': What reflections had participants made in the time since workshop no. 1? Had they had time to discuss the themes with your colleagues? If so, what kinds of reactions had they noted? Then there was a lecture on a theme that the group asked for at the first workshop. After the lecture (30 minutes), we took 15 minutes for questions, comments, and discussion. In this part, the input was a live lecture plus two short video lectures on the theme/approach that the group requested at the end of the first day.

- *Group work.* They split into smaller groups and discussed the perspectives so far. What do you agree on, what do you disagree on?
- *Plenary discussion.* Should society (or can it) regulate and steer the use of nanotechnologies for improving human properties. How could we distribute our perspectives and results to other actors? Who should be informed? How should we develop the themes further in our organizations? The purpose of today's group work and plenary discussion is to formulate a text or a statement. Either one we all might agree on or one where we agree on what we disagree on! This might be formulated as questions, suggestions, warnings, or endorsement.
- *Evaluation of the workshops.* Did they live up to your expectations? Similar workshops will be arranged in five European countries — what might be repeated from this set, what should be changed? If you could decide; what theme(s) should they address?

Participants' influence over the content and organization of the workshops is regarded as a key to their success.

First evaluation

Overall, we were pleased with the quality and the number of stakeholders, the material, and the input for the deliberation. The participants were satisfied as well. The arrangement went smoothly.

The topic chosen for the workshops caught on, SIFO received positive feedback, and people were interested. There was good attendance on both workshops. The participants were well prepared and concerned, and the discussions were good. The presentations were good, and the examples we had picked and presented on the posters were well received and they more than filled their function.

We found participants to be surprisingly reflective and reflexive, and they did make their own choices. The group gradually became aware that it tended towards a perhaps too obvious consensus, based on a 'middle-of-the-road' technological skepticism. So, at the end of the first workshop, they asked specifically for presentations or input that could challenge their consensus. More precise, they asked for perspectives and presentations from enthusiastic trans-humanists, researchers, and social thinkers who applauded the possibilities in nano-based enhancement medicine for the healthy. Interestingly, just days after the first workshop, a Norwegian newspaper had a feature article of a trans-humanist movement within the Norwegian Humanist Organization. The leader of the Humanist Organization could not respond properly because she did not know anything about transhumanism. We wanted to invite the leader of this movement, but he was unfortunately unable to come to the second workshop. The wish from the participants was met by making the participants VJs (Video Jockeys). Here, we understand this term as one would understand DJ — Disc Jockey. DJs play their selections of songs to an audience. The participants twice selected video presentations from an initial selection made by the arrangers.

Just days ahead of the first workshop, we were informed that the lecturer from the University of Oslo was unable to attend the first conference. We checked YouTube, with an eye for relevant TED Talks[5] on the subject. Five presentations were selected. They were presented to the participants in the form short written minutes of each presentation. Interestingly, even though we communicated (quite unwillingly) which one we would suggest they pick, the participants choose otherwise. This goes to show that we had succeeded in communicating to the workshop participants that we wanted them to participate actively. As stated above, there was a specific request from the workshop participants that they, for the second workshop, would have a lecturer that was more techno-optimistic, more than the position the participants felt they were about to take. We were not able to get an appointment, and so again we made the participants into VJs: we prepared a selection of presentations on the subject of transhumanism that they could choose from.

There was some luck in the recruitment. SIFO had struggled to have a representative from the Norwegian Medical Association (NMA) attending and had registered serious interest from the Younger Doctors part of the Association, but neither could make it due to other commitments. However, it turned out that one of the

---

[5]TED is a nonprofit devoted to spreading ideas, usually in the form of short powerful talks. TED began in 1984 as a conference where technology, entertainment, and design converged, and today it covers almost all topics — from science to business to global issues — in more than 100 languages (https://www.ted.com/about/our-organization).

representatives from the User Committee was in the Board of the Older Doctors' Associations.

Did everything work out according to our plan? Did anything go wrong? 'Wrong' appears to be a strong word, and in a sense nothing went wrong. However, some things obviously could have gone better. Recruitment was slow at first, and some organizations we wanted to attend declined, such as the earlier mentioned Norwegian Medical Association and the Medical Ethics Council. In addition, some of the participants were only present on one of the workshops due to other appointments. Some flexibility was demanded from our side: The presenter from Faculty of Dentistry had a PhD defense that was moved to the date for the first workshop. This was solved by making participants VJs as described above.

At the end of the first workshop, participants saw it as a problem that there was a bit too much agreement/consensus in the group; that they got stuck in a rather common positive but worried position. Because of that, they asked to get some input from enthusiastic trans-humanists, just to widen the perspective. But even after two very positive video lectures and a live lecture from the cutting edge of nano enhanced odontology, they still complained that their perspectives had not developed as much as they should have during the process. As arrangers, we felt that this self-reflexivity to a degree contradicted their statement and that this awareness of the situation and of the problems of making statements in the field shows that the deliberation had succeeded in going one step beyond the obvious.

Analysis: novelty

As mentioned, we were aware that this type of workshop always runs the risk of coming up with rather unoriginal and predictable recommendations or statements from the group. A replication of earlier outcomes would not necessarily mean that the 3GDP methodology was a failure because simulating a societal dialogue and initiating debates in the organizations that typically will be asked to give comments and advice to future hearings is obviously meaningful, even if it is perhaps more 'outreach' than 'dialogue' and 'co-creation'. Nevertheless, we wanted to try to create a setting where the potential for new outcomes was increased. To achieve this, we relied on the following three measures or methods:

1. First, there was the idea that more specific themes and questions might provoke some unexpected responses.

2. In addition, we were very ambitious with the workshop organization, the content of the material and posters, the design of them, quality of presentations, time, and space for debates, and so on.

3. Finally, we introduced some elements of workshop democracy, where participants decided on what themes to highlight on the second day and to choose between specific video presentations that we suggested on the chosen theme.

It is, however, necessary that we reflect on what degree of novelty, originality, and 'newness' we realistically are entitled to expect. Researchers have been reading, writing, and practising in this field for years, and they know what kinds of ideas have come forth from other groups all over Europe, Australia, and the United States (Strandbakken, Scholl, and Stø, 2013). This means that we know when the group is touching on original views and when they are just replicating the standard approaches. But it is probably unrealistic to expect a group of stakeholders to come up with a set of never before thought of ideas for future science policy and for medical ethics. We were not anticipating a paradigmatic revolution; we were trying to create a fruitful setting for discussing the interface between medical science and technology and society, society represented by a group of central stakeholders, where we hoped to observe some incremental development of positions, statements, and worries.

Political impact

Even if the project was financed by the European Commission, the deliberation was not on a specific assignment from a central political body. Accordingly, it was not able to guarantee that the group's statements would influence policy directly. So, the problem of unmet expectations is addressed more indirectly. We believe that the stakeholders and their 'constituencies' (organizations and institutions) should be regarded as political actors, not the least in the perspectives of New Governance. In Norway, these organizations and institutions are among those that will be included in future hearings, in state sponsored ad hoc groups, and so on, meaning that they will play a part in decision-making and policymaking on a societal level, however indirect. Further, they will be central in all sorts of societal dialogues on these themes. We anticipated both internal dialogues and processes in the stakeholder organizations and — in a second step — more extrovert activity from them, and we did observe some activity in these directions from some of our participants.

Workshop democracy

In 3GDP, we try to let participants influence the flow of events, the choice of (sub)themes, and the types of expertise needed. This is because, ideally, we want the group to formulate genuinely original perspectives and statements, obviously within the overall chosen theme (here nanomedicine and human enhancement) and the sets of dilemmas. This means that we will encourage group democracy, even if the arrangers will have to provide a programme and some content if deliberations come to a halt.

In the Norwegian workshops, the democratic element appeared immediately. The group exerted its authority both by choosing other presentations than the ones the arranger suggested and by asking for a more techno-optimistic set of perspectives to be presented in the second workshop.

Outcomes/statements

The following is a condensation of the statements from the two subgroups that presented their views at the end of the second day of deliberation. The condensation is based on both written statements and oral add-ons from the presentation and the subsequent debate. This version of the workshop outcome is formulated and written out by SIFO in a secretarial role and later approved by the group as a correct representation of their views. All the material comes from what participants wrote or said.

Later, we comment on the outcome, its relevance, and originality. But we present it here in an unedited context, where the unoriginal and plain perspectives are placed alongside with more interesting ones:

STATEMENT FROM DELIBERATION ON NANOMEDICINE AND HUMAN ENHANCEMENT

Regulatory measures, freedom of research, research as problem solving for society

The main task of the public health services is to treat or prevent illness, even if we are aware that definitions of illness and health might be problematic.

The technology will probably develop anyway, but we believe that national rules and bans could have a normative signaling effect, both inside national communities and between them. We know from experience that in hearings, NGOs regularly refer to what is done in other countries. NGOs often have a rather intensive international cooperation.

For medicine, a paradox of development was presented. Publicly financed Norwegian research is very strictly regulated and very oriented to specific cures and treatments. The element of free and creative research is very small, even if the Research Council talks about freedom of research. This is one reason why Norwegian research often appears to be uncreative, predictable and underachieving. In addition, the criteria for approval of a new drug are so strict with respect to animal testing, informed patient consensus, reporting to the Norwegian Data Inspectorate etc. that much potentially important and groundbreaking research is stopped in advance. On the other hand, it is not very tempting to reduce the ethical requirements.

Commercial medical research, or drug development, on the other hand, tends to cater only for the large groups of patients, because 'minority medicine' does not have a potential market large enough to justify the enormous investments. The question: how many suffer from this disease or problem becomes crucial.

These two constraints on public and commercial science/product development respectively, tend to qualify the above statement that technology will develop anyway. It also leads to too steered research, politically or commercially. We suffer from far too little 'risk taking' research, free research and surprising innovations, which is a bit strange, since we (Norway) could afford to do otherwise.

Governments should regulate and define what health services and what lines of research that is to be publicly financed. The government has to continue using economic incentives to secure ethically sound and responsible prioritizing in research and treatment.

The main principle is that research should be as free as possible, within a set of defined ethical rules. Publicly financed medical services are to be reserved for curing of illness.

Big private players (medicine, agriculture) influence the development and diffusion of technology.

Society and societal debate

It is extremely important that such debates take place, but it is hard to arrange successful public debates because people tend to shy away from this type of ethical problems (prioritizing in life or death situations etc.).

What should be done to facilitate fruitful public debates?

- Ensure high quality dissemination of research and research findings
- Make sure that it is understood that everybody has a right to participate in these debates, there are no definitive answers and you do not have to be an expert to take part — Organizations and institutions should debate the matters internally, at all levels - Debates should utilize traditional and newer social media, simultaneously

It is important that citizens are concerned with and take a stand in these matters because

- If the ethical 'answers' are not in place immediately they have to be adjusted. All perspectives are valuable in this process. A broad participation is crucial when we aim at good rules and solutions — It is also important that the population experience real influence over matters; that important things are not decided upon and develop beyond citizens' sphere of influence
- The population should be sufficiently informed to be able to take responsibility for their own health. That said we need to have a continuous debate over how much responsibility should rest with the individual.

Feedback from participants

Until the closing: At the end of Day 2, we asked, in a rather general way, if the participants were satisfied with the workshops and if they had found it worthwhile. It seemed as if the immediate reactions were overwhelmingly positive. Participants were enthusiastic about the themes as well as about the method.

Post-event: Just days after the final workshop, we received a handwritten note from one of the participants [our translation]:

Hi XXX. Thank you for two nice workshops at SIFO which has given me many thoughts on the healthcare system of the future. I hope you can visit us in the User Committee for Health South-East some time during autumn. I have suggested YYY [the presenter from University of Oslo, ed.] as a presenter at the yearly meeting of the Older Doctors' Association in November.

This later materialized, and one of the SIFO arrangers was invited to speak at a seminar arranged by the User Committee on personalized medicine in December 2014. After the seminar, the User Committee published their position on the subject, partly based on the presentations given [our translation]:

The User Committee asks Heath South-East RHF in cooperation with national authorities to work towards an open and democratic process for the adoption of personalized medicine so that patients in the prioritized areas get accurate diagnoses, targeted treatment, and better follow-up.

Comments on novelty and relevance

Parts of the human enhancement group's statement do go beyond the obvious: Their observations on normative 'soft law' within and between nations are interesting. So is the rather sophisticated analysis of how publicly and privately financed ('commercial') medical research both fail to deliver optimally in relation to societal needs. The dilemma between necessary or fruitful risk-taking on the one hand and a need for ethical regulation on the other hand is real. In addition, the level of self-reflexivity seems to be something new, compared to the deliberations we have reviewed earlier. Even if this reflexivity is not directly translatable to RRI, it probably makes it possible to start citizen and stakeholder involvement on a higher and more realistic level.

We saw, however, that while the deliberation content as well as the debate was concerned with the specifics of bio-nanomedicine and Human Enhancement, the statements at the end of the deliberation *tended towards the general*, even if the themes are understood and implicit in most of the statements. This was a bit disappointing; the specificity of the deliberation tended to disappear from the statements, even if the general statements remain insightful and interesting.

The comments to and statements on social debates are less original. This does not mean that they are superfluous and unnecessary. For the involved stakeholders, this was an important step in clarifying their own perspectives on the role of social debates and on their respective organizations' role in these processes.

14.4   Some Observations on the European Events

As it turned out, the Norwegian deliberations never really worked as a pilot because the European task participants were given too free hands by the consortium leaders. In the absence of a common theme and a common design, it is not possible to include the European events in an overall comparison. We try, however, to use some details from the different events to shed some light on the method — both where they (partly) used the Norwegian model *and* where they deviated from it. Owing to the lack of coherence and structure in the overall task, we present experiences from the European deliberations in a rather loose and fragmented way.

Novel or unexpected results

Germany: For the University of Stuttgart (USTUTT), one of the most interesting features of their attempt at a third generation deliberation was the involvement of such professional groups, users, and consumers that previously had had little or no access to the societal discussions on nanotechnologies. Their theme was new, *nano-based construction technologies*. It was hoped that such a group of participants would develop and present novel statements. As the role of more 'professional' stakeholders such as builders and real estate developers was kept limited, painters, plasterers, architects, and other end users as well as consumers had the chance to come up with original arguments that reflected the preferences of their professional groups. One very precise and specific response was the idea that it is *not interesting to the user if it is nano or not*. The important thing is if it works or not, and what upsides and downsides it has (user-friendliness, toxicity etc.). All new construction technologies and products should be evaluated together under the same premises of benefits and risks.

The Netherlands: The deliberation was on *nanotechnology and food*, in particular on how nanotechnology applied to food can *contribute to health improvement*. Here, stakeholders complained over too much replication of previous results. Some participants who had been involved with the topic for a longer time indicated that they did not hear novel perspectives. They did not clearly define this as a disappointment, but there was a feeling that it will be necessary to introduce some changes in future discussions. If not, the discussion will end. Parties seem to be tired of a discussion that does not go forward. This is an interesting conclusion and one that has bearings on questions of sociopolitical impact as well.

Austria: The focus was set on current and future applications and innovation processes in nanotechnologies — sensor technology, nano toxicology and health, nanomedicine, and safety. Participants worked in small groups on specific topics, leading to the effect that more concrete opinions have been formulated, new contacts were established, and targeted plans for the future of nanotechnologies in Austria have been forged. It is important to bring science, technology, and society in harmony to create a convergence. This is based on the school education. The training must keep up with the enormous development of our time and ensure that qualified staff for the future should take up new task. It is important to actively support social analysis of the use of new technologies, particularly of nanotechnologies in the consumer-related area.

Italy: Issues related to food quality are highly relevant in the public agenda, but the topic of nanotech and food is relatively new for the national context. On the one hand, this aspect helped create a new and original deliberative process. On the other hand, it challenged participation and engagement of stakeholders. Most of the players in the food sector do not consider nanotech as relevant for their business practices and thus have a limited interest in the topic. These issues have been addressed through a careful preparatory work to identify and engage interested stakeholders (in the food sector) in the 3GDP and design with them the scope and structure of the initiative. The Italian deliberation was successful in providing an opportunity for dialogue amongst stakeholders. There has been an ample participation, from different stakeholder groups, though some stakeholder categories were likely under-represented (particularly consumers' organizations and some of the players of the food value chain). The participants agreed on several of the issues discussed. They suggested pursuing a responsible development of nanotechnologies in the food sector. The novelty and originality of the deliberation process was mainly the format of the initiative, the chosen topics, and the type of stakeholders involved in the discussion.

France: The novelty aspect remains unclear. One weakness of the French workshops may be that the topic of improved photovoltaics through nanotechnology is rather uncontroversial; we all want more effective solar panels.

In Germany, questions of to what extent implementation of nano actually would work and whether it would do harm emerged as important considerations. In Italy, the national food industry player claimed nano solutions were not yet implemented, even if such implementation of nanocomponents had been discussed in other countries. In the Netherlands, the participants felt there were no novel insights or arguments that emerged on the debate of nanotechnology and food. However, there was an issue on the balancing of risks and benefits. Some participants mentioned that health benefits as a topic was surprising: they were used to only think of negative health effects and risks. In Austria, the participants voiced a need for an open and transparent communication about the potentials of (bio) nanotechnology, especially for human health.

Norway and Germany experienced some novelty in content; there we saw substantial responses to the nano challenges. The Netherlands looked for, but failed to experience any novelty, while Austria, Italy, and France mainly report to the novelty of the 3GDP method.

Political impact

Our focus is on what happened to the participants during the deliberation, in addition to what we believed happened and could have happened in the aftermath. Political impact might be achieved in a lot of different ways. We could have an assignment from national or local political bodies, we might influence important stakeholders, or we might initiate media debates and so on. In an earlier European project, we had struggled with a concept of political impact that we felt was too narrow. To repeat the initial analysis: In the governance perspective, it is relevant to regard stakeholders' 'constituencies' (businesses, CEOs) as political actors, and hence the stakeholders' representatives might be seen as policymakers, not the least in the perspectives of New Governance. Political influence is about much more than the accumulation of votes in democratic elections. Even if this might tempt us to make it too easy to report political impact.

For Germany, the event and its results might have had some indirect political impact. The representatives of the Baden-Wuerttemberg State Ministry for Rural Areas and Consumer Protection, with its own NanoDialogue Baden-Wuerttemberg, said that they would take up the results on their Nano-Portal. This might also influence the Ministry's work, both on state level and in federal bodies. Furthermore, the guest experts from RKW Kompetenzzentrum and BG BAU had the possibility to bring the deliberation experiences forward as stakeholders in different networks and bodies. Although the political leverage of the professional end user or student participants might be limited, the awareness that the deliberation created could lead to discussions on similar questions in the different guilds, chambers, or other professional associations.

The Netherlands: IVAM invited various stakeholders from Ministries, but they did not attend. However, there were participants present from multiple affiliated organizations that directly advise and inform the Ministries. The deliberation partner RIVM is an example of such an organization. RIVM is the Netherlands' main public sector knowledge institute in the field of public health, nutrition, safety, and environmental management. It conducts research and has practical tasks that are intended to promote public health and ensure a clean and safe environment. RIVM's main commissioning clients are the Ministry of Health, Welfare and Sport (VWS) and the Ministry of Infrastructure and the Environment (I&M), as well as their respective Inspectorates. RIVM and the other affiliated organizations, like the Food and Consumer Product Safety Authority (NVWA), are close to policymakers and policymaking. Moreover, many participants are part of various consultation bodies on nanotechnology that are of some stature.

Such advisory bodies consisting of multiple stakeholders are common in the Netherlands. There is agreement on, for example, their formal status and on the reliability of their advice. Since their role is formalized, their impact is bigger than other stakeholder discussions. Further, the results of the deliberation process will be spread among policymakers and these consultation bodies and can have an impact that way. Even if the Dutch stakeholders felt that the deliberation did not produce any novel insights, it appeared as if they did see it as part of a *continuous societal debate* over questions of nano in food and health issues.

Austria: The Federal Ministry of Health submitted a statement on the Austrian nanotechnology initiatives and the importance of the key enabling technology in the country and asked for the outcomes of the discussions for further dialogues within the Austrian Nano Information Commission (NIK), where all nano relevant Austrian Ministries are represented. A report to be submitted to the ministries was prepared (in German). Beyond that, political impact might result from increased reflexivity among scientific and business representatives.

In Italy, participants agreed that a clear view of nanotech opportunities and challenges for the food sector is lacking and there is a need for more information at all levels, both for experts and business operators and for the general public. The 3GDP was considered useful to address this information gap and to increase the nano attention of the players in the food value chain. The wide range of stakeholders involved in the events was helpful to ensure that different visions and opinions were represented in the debate and stimulate an open discussion. Participants were very interested in gathering the views of the different stakeholders participating in the workshop. The event was successful in stimulating reflection on the responsible development of nanotech and fostering a dialogue amongst the various stakeholders, pointing out some relevant points and needs for responsible innovation. This was likely the main impact of the 3GDP. Potentially important for the follow-up phase was the presence at the workshop of journalists. At least a couple of articles on food in the sectorial press were expected based on the talks and documents presented.

France: Stakeholders from different levels of government participated in the deliberation; there were mayors and elected officials from nearby cities of Grenoble and representatives of regional governmental institutions. They all have a stake in questions about the role that nanotechnology innovations can bring to the citizens as challenges of the twenty-first century; their participation was a factor in the success of the deliberations and their viewpoints have provided original thoughts that had not been discussed so far.

Workshop democracy

For the third generation deliberative processes, we want to let participants influence the flow of events, the choice of (sub) themes, and the choice of expertise. In this way, we hope to increase participants' feeling of ownership to the processes. In addition, this is a means for making the deliberations more relevant and interesting for the participants. We want the group to formulate as original statements as possible, within the overall chosen theme and the sets of dilemmas.

Workshop democracy was given room in the German Deliberation too, even if the USTUTT team planned and structured the event: In the beginning of the event, the participants continued their discussion with the guest experts as the planned Tour de table was skipped. Furthermore, the participants were given the chance to decide on the themes of the dialogue stations — after suggestions by the USTUTT team — and thus influence the questions and aspects that were discussed.

In the Dutch deliberation, participants influenced the flow of events in multiple ways. During the workshop, participants decided on what to discuss or which topic to use. The participants of the first day defined the approach of the second day: they provided possible solutions, topics, and 'research questions'. A most interesting example of workshop democracy is the decision to hold the workshop under the Chatham House rules, freeing participants from having to represent the views of their constituencies. Anyone who comes to the meeting is free to use information from the discussion but is not allowed to reveal who made any comment. This is seen as the way to discuss controversial themes more freely and less constrained by obligations and loyalties.

In Austria, participants were asked to introduce their field of expertise and the related challenges they are facing. New contacts have been established in order to work together synergistically on new technology challenges.

In Italy, the deliberation was designed together with stakeholders active and interested in the theme. The final topic and structure of the day was defined with them. Several participants to the workshop provided comments and input during the various phases of the process, but for many other participants, the event was mainly informative and their interest (and capacity) in providing input during and after the event was limited.

The French workshop started in an informal way by visiting the R&D Labs so that it offered the opportunity for the stakeholders to speak directly with the researchers. Thus, the discussion started with the researchers, so that every stakeholder felt free to discuss whatever he wanted.

14.5 A Future for Third Generation Deliberations on New Technologies?

There appears to be a growing reflexivity on the relationship between technology and society. This was most pronounced in the Norwegian 3GDP exercise: here, the stakeholders came up with new arrangement for sponsoring 'critical' research.

We believe that both the governance aspects of stakeholder workshops and the participatory democracy aspects of citizens' workshops will benefit from the more elaborate design of the third generation deliberations. The six national events all were regarded as successful by arrangers (NanoDiode WP 3, Task 1 consortium members) and by participating stakeholders. The methods and the approach were well thought out and they seem to be transferable across national cultures. This makes it possible to offer some overall evaluations and conclusions.

The 3GDPs have been conceptualized based on experience with and analysis of such processes since 2004. The problems addressed by Stø and colleagues (Strandbakken, Scholl, & Stø, 2013) are real problems based on observations of real events. So, it was not a surprise that the method works.

The generational perspective implies that things develop and that they improve over time. This means that even if we regard the French Île-de-France deliberation in 06/07 as a model of how citizens' deliberation should be done with respect to resources (institutional resources and time) made available to the group and with regard to post-event reporting, we still believed that we could improve on it. This is because we had the benefit of afterthought. In our view, this deliberation tended to take a too wide set of themes ('nano and society'), and it was a bit unclear on political impact, mainly because of the difference between regional and national levels of government.

However, our possibility of making a sound assessment of the method suffers from the deviation from the model. This concerns the number of days set aside for the exercise (one or two days), the time between these two days, the number of participants involved, which stakeholders were involved, and the themes. Because of these deviations, we build our assessment on the future of third generation deliberative processes mainly on the Norwegian event.

Design

A third generation deliberation on nano or other emerging technologies should follow a structure where (pre-recruited) stakeholders or citizens agree to participate in a two-workshop event. Before the first workshop, participants will receive good quality material on the theme/topic of the deliberation and on what a deliberation is and what is expected of them. The workshop should be professionally moderated/organized, and lectures and material presented should be of high quality. Participant should be encouraged to try to influence the choice of sub-themes, suggest lecturers or lecturers for the second workshop, and so on.

The second workshop should be arranged a week or two after the first one — for participants to be able to discuss the themes with their organizations or with their families and neighbourhoods, to give them time to reflect, and to give them the opportunity to seek out alternative information. At the end of the second workshop, the group should issue a statement — either as the view of all (consensus) or as a balanced presentation of differences.

How the event is situated in the overall political system should be clear from the outset. Sometimes it will be producing specific input to legislation and regulation, but the decisions are to be made in the representative political system. If possible, try to generate some public or media attention around the workshops, a main task might be to simulate the societal debate that fails to materialize elsewhere.

The group size should ideally be between 10 and 25. The composition of the group must be thought through: similar or different stakeholders, the same for citizen groups, gender balanced or one gender groups, ethnic composition, age, levels of education, and income levels. This depends on what questions to address, but it also has bearings on how results should be analysed and used.

A third generation deliberative process is a useful tool for engaging citizens or stakeholders in democratization of science and technology.

Reference

Strandbakken P, Scholl G, Stø E (2013) *Consumers and Nanotechnology: Deliberative Processes and Methodologies*, Jenny Stanford Publishing (formerly Pan Stanford Publishing), Singapore.

Chapter 15

Participatory Democracy: Hybrid Forums and Deliberative Processes as Methodological Tools

Virginie Amilien,[a] Barbara Tocco,[b] and Pål Strandbakken[a]
[a]SIFO, Consumption Research Norway, Oslo Metropolitan University, Postboks 4, St. Olavs plass 0139 Oslo, Norway
[b]Newcastle University, Newcastle upon Tyne NE1 7RU, United Kingdom
viram@oslomet.no

15.1 Two Approaches

As pointed out in Chapter 1, there is at present a huge 'market' for deliberation and engagement techniques, methods, designs, and approaches, not the least because of the insight that engagement is probably *the* central element in successful RRI. We are confident, however, that our approach in this book — the third generation deliberative process (3GDP), building on the Danish Consensus Conference, the UK Citizen Jury plus our overview of European and American events — remains a fruitful and effective way of involving citizens and stakeholders in debates over emerging technologies.

Consumers and Nanotechnology: Deliberative Processes and Methodologies
Edited by Pål Strandbakken, Gerd Scholl, and Eivind Stø
Copyright © 2021 Jenny Stanford Publishing Pte. Ltd.
ISBN 978-981-4877-61-9 (Hardcover), 978-1-003-15985-8 (eBook)
www.jennystanford.com

That is, when 'we' — the government, the scientific community, a CSO, and so on — want to have citizen or stakeholder views on specific questions, applications, or technological dilemmas. 3GDP (third generation deliberation) is a method for eliciting statements, opinions, or reflections on *predefined* themes. It also has to have a rather clear link to policymakers and/or political processes. If not, it will seem to be merely another focus group exercise.

In the aforementioned deliberative democracy or public engagement market, one approach stands out as a very interesting contrast to the somewhat rigid third generation deliberations. Presented by Callon, Lascoumes, and Barthe in *Acting in an Uncertain World* [French original (Callon, Lascoumes, & Barthe, 2001), English translation (Callon, Lascoumes, & Barthe, 2009)] as a democratic and dynamic way to think and act together, *hybrid forums* can be described as public dialogues with the aim of constructing a common project around a defined challenge. Result and practice oriented in other ways than deliberative processes, and much more positive to controversies and disagreement, hybrid forums might stimulate concrete democratic results and effective dialogues amongst all actors in the value chain. In line with this way of thinking about public engagement methods, we had an opportunity to put into practice hybrid forums within the collaborative framework of the EU H2020 Strength2Food (S2F) research and innovation project.[1] We developed an experimental public engagement method called Hybrid Forum 2.0 (HF 2.0) aiming at shaping dialogues *on food quality schemes and food sustainability in the European Union*.

Although the authors' hands-on experience with HF 2.0 is not based on nanotechnologies and comes from a project about local food and short food chains, the subtitle of the main source of inspiration (Callon, Lascoumes, & Barthe, 2009) is '*An Essay on Technical Democracy*', opening towards many direct links to scientific innovation. Following Callon and his colleagues, HF 2.0 has been developed to bring the traditionally separated fields of research on common sense and 'legitimate' science together under the 'hybrid forum' umbrella of local representations and

---

[1]Strength2Food is a European Union's Horizon 2020 research and innovation programme focusing on 'Strengthening European Food Chain Sustainability by Quality and Procurement Policy'. For more information, refer to https://www.strength2food.eu/

use of sustainable resources and contribute '*to the formation of networks of actors sharing a collective project*' (Callon, Lascoumes, & Barthe, 2009, p. 34). This type of deliberative process involves not only the participation of chosen participants from the planning to the dissemination of results, through exchanges and common experiments, but also ideally *the transformation of perceptions and standpoints in the long-time perspective* through several HF 2.0 forums. Furthermore, the hybrid forum is not only a methodology to perform public dialogues and involve several actors in the field, but also a way to collect data on a given case study as well as a dynamic mechanism for democratic communication and concrete results for consumers.

This chapter presents the HF 2.0 as a new methodological tool for multi-actor fieldwork in social sciences. The HF 2.0 has been described in a previous publication (Amilien, Tocco, & Strandbakken, 2019), but the aim of this chapter is to show and analyse how HF 2.0 differs from the more conventional methods for citizen–stakeholder participation and democratic involvement. In order to do this, we contrast HF 2.0 with third generation deliberative processes. After a first part presenting HF 2.0 as an innovative path to technical democracy, the chapter proposes a strong empirical approach based on experiences from HF 2.0 cases in seven European countries. The third part will mainly focus on results and compare with the 3GDP before we conclude about the potential value of HF 2.0 as a methodological tool in social sciences and particularly in the field of new technologies, innovations, and research development.

15.2 An Approach to Technical Democracy: A Path to HF 2.0

In the beginning was the word. The famous first sentence of the Gospel of John is pertinent here for several reasons. First, because he actually used the Greek word 'logos', and not word, in the original text; logos, which means discourse and reason (or discipline, as in biology and sociology), even reasoned discourse, more than a 'word', is central in hybrid forums. Second, because, following John, all things are made from the logos. While for Aristotle, logos is but *one of the three* methods of persuasion, concerning the proof provided by the speech itself, together with *pathos* — putting the audience into a certain frame of mind — and *ethos* — building on the personal character of the speaker. Based on facts, consistency, and a way of describing reality, logos arguments for a truth, or an apparent truth as specified by Aristotle himself, and emphasizes a philosophical perspective leading us to dialogue and discussion, which also play a major role in hybrid forums. Third, the reference to ancient Greece makes relevant the etymology of the word 'democracy' (from 'demos' meaning *people* and 'cracy' meaning *power*), which is definitively a central pillar of the hybrid forum. So, *In the beginning was the word* is a simple sentence reflecting the main pillars of the hybrid forum. Words are at the heart of dialogues outlining technical democracy.

In deliberative processes, words from experts, from stakeholders, and from lay people are crucial, but in a hybrid forum, they are of equal importance. In their essay for a technical democracy, Callon and his colleagues (2009) proposed to use the challenges linked to scientific knowledge *and* knowledge from local actors in a dialogic democratic perspective that permits to shape a common world. Callon and his colleagues describe hybrid forums *as open spaces where heterogeneous groups can come together to discuss questions at different levels* (Callon, Lascoumes, & Barthe, 2009, p. 18).

A hybrid forum approach should map human resources within the local community in order to identify the mechanisms and drivers that can stimulate innovations and to construct together, as Callon, Lascoumes, & Barthe (2009) put it:

> By trial and error and progressive reconfigurations of problems and identities, socio-technical controversies tend to bring about a common world that is not just habitable but also livable and living, not closed on itself but open to new explorations and learning processes. What is at stake for the actors is not just expressing oneself or exchanging ideas, or even making compromises; it is not only reacting but constructing (Callon, Lascoumes, & Barthe, 2009, p. 35).

One objective of the hybrid forum is to create awareness and facilitate collective exploration and learning, cooperation, and integration of a plurality of points of view. Therefore, hybrid forums are more process and practice oriented than deliberative processes and less based on results.

Dialogic democracy is a key part of social and humanist research. Let us bear in mind that Callon, Lascoumes, & Barthe mean that we definitively need a complementary approach to combine 'outdoor' and 'indoor' research. Indoor research describes a sort of 'closed research world', a form of research where researchers communicate with researchers — the more or less normal conference situation. The classical process for indoor research is explained by Callon and his colleagues in three phases called 'translations':

1. From macrocosm (the world at large) to microcosm (laboratory); the researcher translates the world in a sample, often conceived as a representative one but nevertheless not real as it is a created sample

2. The study of the simplified samples of the macrocosm and the creation of data/scientific data. This phase contains modeling, statistic approaches, and so on that are a part of the researcher's knowledge, interests, and background (a second translation from the real world)

3. Translating the results of the study back to macrocosm/the world at large, quite often influenced by social, economic, or scientific interests, and so not really to the world at large, but to an adapted version of it

Outdoor research is described as a type of research including non-researchers ('profans'). This can be done through an indirect implication (as empirical data) or direct participation. Several methods are currently used, giving an outdoor perspective to consumption research, from the traditional surveys or focus group to the more recent deliberative process. While the focus groups aim at having a common discussion to have better knowledge on one given theme, the actors in the hybrid forums will '*not just expressing (them)self or exchanging ideas, or even making compromises*' but also discovering, learning, and constructing together as underlined before.

15.3   Path from Hybrid Forum to HF 2.0

We present here a new generation of hybrid forums through the experimental method of 'HF 2.0'. Hybrid Forum 2.0 has been developed within Strength2food, and so there is no literature on the approach beyond the project. However, Farías (2016) provides an excellent overview of these methods, and their role as participatory devices, organized in a Chilean city in 2010 (based on Callon, Lascoumes, & Barthe, 2001). Studying and evaluating concrete hybrid forums, the author emphasizes the issue of equality of participation as well as challenges when issues such as power of economy or political decisions took over dialogic democracy — for example, if hybrid forum did not have a collective dynamic approach to controversies (Farías, 2016). Furthermore, a previous article on HF 2.0 gives details about the method (Amilien, Tocco, & Strandbakken, 2019). Note that we refer directly to the article in BFJ and provide here only a short presentation.

Developing Hybrid Forum 2.0 in the Strength2Food project aimed at organizing democratic dialogues on given controversies.[2] We made an intern protocol inspired from Callon and his colleagues, based on concrete actions and values as to find a controversy, to create a public room to have a dialogue for constructing a common project and to observe equality, transparency, and traceability, as well as the clarity of rules while organizing the forums.

Inspired by the third generation deliberative processes, H.F. 2.0 builds on two complementary parts to assure an equal and fruitful dialogue in the public meeting. The preparatory workshop (part 1) aims at reaching a shared frame based on common information on the subject, presentation of the public dialogue structure and rules, and better knowledge about each other's ideas and perspectives. The workshop is then followed by the public meeting (part 2), an open dialogue where everybody can contribute.

The two interconnected parts are organized on the same day, following each other, when it is possible. The preparatory workshop in part 1 is limited to a group of 'Invited Participants', including local stakeholders, scientific experts, local entrepreneurs, and lay people/informants — who are given the possibility to be sitting in the panel. Part 2 is the public meeting, beginning with a short panel discussion to frame the subject and then opening up to the public from the whole community. Participants in HF 2.0 are both speaking and listening, learning (learning by doing and learning from each

---

[2]This page is strongly referring to the article in BFJ and partly reproduced with the authorization of the editors — more details in Amilien, Tocco, & Strandbakken (2019).

other), acting together, and constructing a better local community, or world, together. As controversies can be considered as a pillar of dialogic democracy (Amilien & Moity-Maïzi, 2019), hybrid forums are spaces where controversies take place, but often to provoke, stimulate, and hopefully change the conflictual situation. This is also the reason why H.F. 2.0 is an open participation method, without any predefined goal and not necessarily working towards a written, unanimous statement from the dialogue or the group.

The purpose of developing HF 2.0, as discussed here, involves addressing local controversies by stimulating mutual reflections, better understanding, and, ideally, an open dialogue on technical, political, and socio-cultural issues linked to the local territory and related food systems. While creating a dynamic and interactive meeting place, HF 2.0 brings together the different worlds, via a web of local 'actants', aimed at a collective co-construction process of knowledge around given controversies. In the context of the Horizon 2020 project Strength2Food, we first envisioned HF 2.0 as communication and information tools, while providing a multi-actor research tool for data collection.

In the following section, we present the first HF 2.0 as case studies in each of the seven countries. The seven case studies are here presented in the chronological order, from 2016 to 2019, as the research teams discussed and learned from each other in the meantime.

15.4   Case Studies[3]

Case study 1, Strength2food, Norway

The first HF 2.0 was conducted in the city of Sandefjord in Norway on 30 August 2016.[4] It was our first experiment and this hybrid

---

[3]These national case studies have been done by Davide Menozzi (University of Parma, Italy), Ratko Bojović (European Training Academy, Serbia), Matthieu Duboys de Labarre (CESAER, AgroSup Dijon, INRAE, Université Bourgogne Franche-Comté), Agata Malak Rawlikowska (Warsaw University of Life Sciences — SGGW, Poland), and Peter Csillag (Eco-Sensus Research and Communication and Corvinus University of Budapest, Hungary).

[4]Details available at https://www.strength2food.eu/2016/08/23/how-is-local-fish-valued-by-the-people-of-sandefjord/

forum also worked as a pilot for the project. This event aimed at discussing 'the value of local fish' in the city of Sandefjord. The choice of the controversy was based on various local controversies about fish quality — fresh/frozen and local shop/supermarket — price, and tradition, which emerged during a fieldwork in June 2016 underlining controversies (Fumel, 2016, p. 43). The preparatory workshop involved five researchers (including a master student) and four panellists: a consumer, a representative from the local County Governor, a fresh food manager from the local grocery store, and a local fishmonger. The first part lasted about 3 hours. It began with information about hybrid forum and presentations by two different experts on the regulation of local fishing and Norwegian fish consumption. This led us to a dialogue where panellists shared their opinions, trained for the panel presentation, and agreed about a concrete proposition to arrange a crab festival at the port to promote local seafood. The second part, consisting of the open forum, lasted for about an hour. Panellists, researchers, and the public audience (limited to five people) had an active discussion emphasizing agreements, conflicts, and uncertainties. The main topics discussed included changes in the local fish stock and criticism of national fisheries regulation. The group thought about strategies to better utilize and manage the existing local resources. Following the discussion in this second part, the crab festival idea was put on the side and a new 'bigger' idea was put forward: the establishment of a maritime museum so that local people and tourists can learn about life in the fjord.

Case study 2, Strength2food, UK

The UK first HF 2.0 was hosted on 21st September 2017 in North Shields, a town in North-East England, renowned for its historic fishing port and busy seafood trade operations. The event, organized by Newcastle University and SME Food Nation, aimed at sharing experiences, learning perspectives, and exchanging knowledge on the theme 'Locally landed seafood on the menu: how do we create a supply chain to achieve this?'. The topic was identified via Strength2Food market research, which highlights limited quantity and variety of locally landed seafood on consumer plates. The discussion thus attempted to shed light on the key barriers behind this trend and opportunities for innovative actions.

The first part of the HF 2.0 consisted of a preparatory workshop, attended by different experts from the local supply chain — specifically, five invited panellists (the FLAG programme officer, the managing director of the Fish Quay, the manager of a local seafood shop and fish box scheme from a nearby coastal town, the head chef from a local restaurant, and a passionate consumer/ food blogger) and respective organizers/researchers from the SME and University. The discussants highlighted numerous supply- and demand-side limitations to justify the lack, or limited efficiency, of local fish supply chains. Among various suggestions, key aspects included educational campaigns and marketing strategies to improve public awareness on localness and sustainability issues, as well as initiatives to foster cooperation in the local fish supply chain.

The second part of the forum opened to the wider public. It was attended by approximately 30 participants and included the general public, industry actors, and representatives from the third sector. The discussion focused predominantly on the demand side to better understand consumer habits and trends around local seafood consumption. The roles of perceptions and (lack of) skills in preparation were identified as key barriers to purchasing and eating a wider variety of fish, including lesser known and locally abundant species. Initiatives that could increase awareness and bring about behavioral change were emphasized, including educational initiatives in school curricula and sustainability campaigns endorsed by celebrity chefs.

Case study 3, Strength2food, Serbia

The first Serbian hybrid forum was organized in Petnica Research Centre, near Valjevo, on 27th November 2017. As a part of a larger controversy about school meals, the HF 2.0 aimed at a common dialogue about improving the quality of children's nutrition in Serbian primary schools. The approach was to give a positive title to the complex controversy we chose for the first HF 2.0, so that participants had sense of contributing to something good, relevant not only for them but for other participants as well and for wider community (so words like 'improving', 'opportunity', 'creating' were in the title).

Invited panellist on the first part of the HF 2.0 were representatives from the Ministry of Education, Science and

Technological Development (2), directors of the local primary schools (2), local agriculture extension service (2), an expert for public food procurement, the farmer, the doctor from a local hospital, and the school chef. Initial presentations led to discussion, which mainly focused on the lack of regulations on nutrition in Serbian elementary schools and challenges with the public procurement, leaving to the schools to find their own ways and mechanisms to improve the quality of school meals. Although it was not an immediate result, the follow-up was defining the topic of the next HF, that is, 'preparation of regulations on food quality standards in Serbian primary schools', which took place at Belgrade in the following year.

The public part gathered 34 participants, with diverse background, interest, and topics (directors and teachers from local primary and secondary schools, nutritionists, health workers, producers, and parents), which brought vivid discussion. Still, one topic emerged as the most important for the whole group — the need to raise awareness about healthy eating of children and their nutritional habits, which are acquired and developed in their families. That led to the creation of questionnaires for children and their parents by EUTA (European Training Academy, Serbia) and MPNTR team. Questionnaires about the children's food preferences, parents' knowledge, attitudes and practise on healthy food, and the children's food diary were implemented during 2018 and 2019.

Case study 4, Strength2food, Italy

The first Italian HF 2.0 was organized in the Appennino Tosco Emiliano in Fivizzano, Massa-Carrara province, on 1st June 2018 as a part of the Sapori festival. The hybrid forum focused on agriculture and food and searched for instruments to counteract the depopulation of the MAB-UNESCO (Man and Biosphere Reserve).

During the first part, experts and panellists met and discussed during a couple of hours. They were from a Local Action Group, an agricultural holding 'Zafferano', the local Slow Food group and the Community Presidium, the owner of the family restaurant, a former mayor, a Senator, and a student in veterinary in Bologna, as well as two agricultural economists and two students from the University of Parma. The discussions focused on existing and potential answers or strategies to face the depopulation of rural areas, which is a huge problem in the studied context. In addition, some points were also drawn representing the difficulties and barriers to be overcome. Getting inspiration from other Italian rural areas and emphasizing the role of the education in promoting the production of food products related to this territory were considered as solutions.

The local public actively participated to the open dialogue in part 2, which lasted around 2 hours. After the panellists had talked, people told about their experiences and opinions. It seemed central to have this HF 2.0 opportunity for mutual listening and the search for a common solution.

Case study 5, Strength2food, Poland

The first HF 2.0 in Poland was conducted on 17th July 2018 in the city of Płońsk, a town in Mazovia region, located 70 km from Warsaw, the capital of Poland. The discussion focused on 'the tradition and future of the farmers' market in Płońsk'. With more than half a century of tradition, the local farmers' market takes place twice a week, on Tuesdays and Fridays, in an open-space square in the heart of Płońsk. However, there are ongoing discussions on whether the market should be moved to a more convenient and larger space, with better parking facilities. Moreover, more investment is necessary to improve its functioning and facilities.

The first part of the HF 2.0 — preparatory workshop — involved four researchers and seven panellists: the city council, including the mayor, representatives of the departments for municipal investment and food markets, the farmers selling on the market, consumers, and residents of neighbouring streets. The first part lasted about 1.5 hours. It began with information about hybrid forums and presentations by two different experts on the results of the research study devoted to the consumer and producer perception of functioning of the farmers' market in Płońsk. Afterwards, the panel had a productive round-table discussion covering all potential aspects of the farmers' market organization and management, identifying suggestions for the future and pointing at investment possibilities and plans.

After a break, the second part of the hybrid forum began, calling in a wider audience (open forum), with about 20 representatives from citizens, farmers, and traders selling on the farmers' market to anyone interested on the topic under discussion. It started with the panellists' short speeches, followed by the discussion with an audience. This part also took about 1.5 hours. Thanks to the discussion in the first part of the forum, the panellists' statements were more consistent with each other and focused on the problem discussed. The debate was very dynamic and many ideas regarding the improvement of the farmers' local market were put forward. The citizens, farmers, and municipality representatives agreed on the fact that the current traditional location of the market is convenient for all parties. However, some investments are needed in order to make it more functional — including possible roof coverage, stands, and, in particular, a larger car park.

Case study 6, Strength2food, France

The first HF 2.0 organized in France took place in the village of Hauterives, in the Drôme department, on the 27th of August 2019. The choice of the theme was co-constructed with the initiator of a market of organic and local producers, called the 'Little Market'. His wish was to think about the possibility of making this place more than just a commercial place by adding an associative and citizen dimension which was not without controversy. Were a commercial activity and a civic activity compatible? What were the links with the local environment (other associations, the municipality, etc.)? And finally, who would be the audiences for this project?

The first phase of the HF 2.0 involved a researcher and a panel of seven participants: a producer who owns the place where the 'small market' is located, two representatives of associations, two elected municipal officials, two citizens who are consumers of the 'Little market', and a guest expert in rural development. This first part lasted about 3.5 hours. The main discussion revolved around the idea that the 'Little market' was a support for an associative activity. Through the discussion, several points emerged that deepened or went beyond the initial project. Among the main ones, we can retain: the question of governance, of audiences, the need for access to 'good food', and the issue of participatory democracy. This second part lasted about 2 hours and brought together 35 people. The panellists organized the space in such a way (no tables, all seated in a circle) to encourage participation. After a brief presentation of the framework by the researcher (who explained to the public the presence of the cameras), the discussion began with a brief reminder of the project, followed by a round-table discussion where everyone was invited to introduce themselves and explain, if they wished, the reasons for their presence. The use of a 'paper board' allowed everyone to keep track of their expression. The discussion went well beyond the scope of the project. Indeed, the debates very quickly turned to general questions about the dominant agricultural and industrial model and the need, or not, to transform it. Owing to the large number of participants and the great freedom of expression, the debates did not necessarily lead to concrete projects and may have seemed a little frustrating. In spite of this, the paper board technique made it possible to identify major trends and above all to agree on the organization of a second hybrid forum.

Case study 7, Strength2food, Hungary

The first hybrid forum held in Hungary was prepared in Szekszárd on 22th October 2019. This event aimed at identifying the importance of regional quality food labelling. In 2011, Eco-Sensus Nonprofit Ltd — the Hungarian partner of the project — had developed a regional food quality label as a coloured figurative certification mark in order to identify and promote local food products in the region. Since then, an online database of 100+ producers had been maintained and several promotion activities were developed.

With this first hybrid forum, the main purpose was to have a multi-stakeholder assessment of the system: whether there was a real need from both the supply and the demand side. The forum included local producers (e.g., winemakers, meat producers, and fruit and vegetable producers), consumer organizations (e.g., local consumer organizations, local restaurants, and public caterers), and several other stakeholders with various activities (e.g., from the local government, the manager of the local farmers' market, and tourist agency).

The starting point was that many food quality labels exist that might confuse the consumers. However, regional food quality labels do have their role to educate and also to ensure consumers are interested in purchasing local foods. In the local farmers' market, it is very difficult to guarantee the origin of the products sold as 'local food'; however, a regional food quality label could serve as a good basis to build the trust in the consumers. Szekszárd is a famous Hungarian wine region, where winemakers have their own traditions and regulations, which could serve as an example for other local foods. Regarding public catering, the public caterer of the municipality Szekszárd is responsible for a daily 1100 portion of food. Sourcing local inputs have several barriers, first of all their technology requires processed and ready-to-cook inputs (e.g., peeled and sliced potato) what majority of the local producers cannot produce. At the same time, constant quality and sufficient quantity are the most common reasons why it is difficult to use local quality inputs in a large-scale catering system. Local restaurants struggle with similar problems; however, for meat and spices, usually they can purchase locally.

Notes about the seven case studies

All Hybrid Forum 2.0 were originally based on a local controversy that the research team defined as pertinent and interesting. But after the first HF 2.0, it would happen that the involved participants were more interested in 'constructing a common project' together than to build a new dialogue on another controversy. This would be a central issue of the second HF 2.0. The potential role of the first HF 2.0 for deliberative democracy will be discussed in the following section, alongside another public engagement method, which is the 3GDP.

15.5   Comparing HF 2.0 with 3GDP

As previously mentioned, the HF 2.0 constructed by and used in Strength2Food are directly inspired by the 3GDP both by their concrete experiences in the field and by the impact of their results. On the contrary, hybrid forums have different objectives and different approaches emphasized both by their 'hybridity' — between private deliberative processes and public debates — and by the objective of 'shaping a common world'.

HF 2.0 involves not only the participation of chosen participants from the planning to the dissemination of results, through exchanges and common experiments, but also the transformation of perceptions and standpoints in the long-time perspective through several hybrid forums. For example, in the Norwegian case study, the value of local fish in a Norwegian city has been chosen as a subject for hybrid forum where all interested actors could meet and exchange knowledge. Panellists and public were/would be able to meet and discuss the subject not only at the first but also second and third hybrid forums for eventually finding the best view and possibly solution for local community. All potentially interested people are invited, including stakeholders and consumers, but also private initiatives or small-scale entrepreneurs that often play a major role in developing local businesses.

Invited participants are told that the hybrid forum essentially aims at building a better world together. They have been told that when they participate to a hybrid forum, the point is not to defend an idea or personal interests, but to share their experience, their knowledge, and their expertise with the group and later together with the public. The HF 2.0 is not served by defending a personal interest but by understanding and preserving the collective interests to develop together a better local environment. Social roles and ready-made ideas have to be put aside to be able to better listen, talk to each other, discuss, converse, and question: To be a part of HF 2.0 is to accept to have an open mind, to talk together, to listen, to share, and 'shape a common world', as Callon *et al.* put it.

The participant mix in HF 2.0 tends to be different from the one in deliberations. HF 2.0, with their heterogeneous groups and opportunity for everyone to be part of the dialogue, are more open to the unexpected, while 3GDPs tend to recruit either citizens/consumers or stakeholders in order to deliver policy relevant input. HF 2.0 streamlines different interest, perspectives, and processes, which take place in wider context, and puts them in an unplanned direction.

Another major difference is inherent to the objectives. HF 2.0 do not build on a pre-defined controversy that it aims at solving. Eventually, the methods are, to a certain extent, also unlike. A part of the HF 2.0 methodology focuses on the controversies that have/will emerge(d), or develop, at the local level, among the local participants and that should not be defined in advance.

This probably holds, even if HF 2.0 has adopted some elements from 3GDPs, such as a clearer structured two days' design or at least two parts' structure. Further, the role of researchers is quite similar for both methodologies in part 1, with a workshop and panellists, but then completely changes in part 2, where researchers organizing HF 2.0 are supposed to be humble and neutral. Their role mostly consists in creating a stable frame around the hybrid forum, organizing a structure that permits and facilitates the dialogic discussion.

Moreover, the most important result of the HF 2.0 is obviously the process around the hybrid forum itself, and not necessarily the political, economic, or social consequences of it. Learning and thinking together is a clue. The objective is not to solve a problem, but to create a dynamic room where outdoor and indoor research can meet by emphasizing a controversy and making local people discuss around it.

The utopian project of shaping a common world together can obviously be concretized by HF 2.0, for example, for assessing the possibility of innovation or the importance of tradition in Geographical Indications and also for reflecting to product development based on nanotechnologies on the everyday market.

A HF 2.0 approach builds on a multidisciplinary and open-minded perspective, which is not far from conceptual engineering: actants from the field are the ones who know and can have a dialogue. The open structure aims at overcoming the limitations of the conventional approaches, as in European research where the project structure sometimes constrains and even pre-defines outcomes. The ambition is to further open up the quite classical research approach to the 'unexpected'. To be able to consolidate research with participating methods in the future, HF 2.0 needs to be dynamic, interactive, opened, and constantly adapted and regenerated by new ideas, controversies, and uncertainties. When HF 2.0 become outdated, a new generation of engagement methods will be required.

15.6 Conclusion

Working on a European proposal about sustainability and food origin for H2020 Strength2Food, our research team developed the HF 2.0 methodology inspired from Callon and his colleagues (Callon, Lascoumes, & Barthe, 2009). Hybrid forums are used in a dynamic and positive way to reach a mutual discovery and comprehension, often towards an uncertainty. Everyone, both specialists and laypersons, contribute to enrich the discussions, which improve knowledge. In this approach, the distinction between facts and values is suppressed. As Callon and his colleagues put it:

Hybrid forums are the cubicles in which existing facts and values are mixed in order to be recomposed and reconfigured (Callon, Lascoumes, & Barthe, 2009, p. 233)

What is essential for ordinary citizens and laypersons in dialogic democracy is not participating, but weighing up and contributing (Callon, Lascoumes, & Barthe, 2009, p. 248)

To summarize, we can say that the purpose of HF 2.0 is to create the opportunity for a dialogue aiming at constructing a common project to improve the local community on a defined 'controversy'. Therefore, HF 2.0 builds on four main pillars: (i) controversies and disagreement, (ii) equality and openness, (iii) transparency and traceability, and (iv) contribution to construct together a common world.

For regular politics, for example, governmental needs for public and stakeholder meetings giving legitimate input to political processes, the 3GDP approach is extremely fruitful (see Chapter 14). The strengths are based on clear questions or problems, a clearly defined role vis-à-vis the political system, and a set of techniques for eliciting responses that go beyond the traditional or obvious, while remaining inside the previously agreed-upon framework. Here lie the advantages of such deliberations as well as their limitations.

For the HF 2.0, however, strengths and limitations are different. Actually, they are partly the opposite, even if the part one of the HF 2.0 seems to be quite similar to the participatory workshop of 3GDP. HF 2.0 builds on a dialogic interaction where different stakeholders and lay people can speak equally and where a free discussion with little direction from the facilitators is ideal. It can then be considered as a more anarchistic phenomenon, and it operates best in less clear environments and when situations are more undefined. The *raison d'être* of HF 2.0 is more about doing than about reporting. Like the Socratic dialogue, the process is mostly part of the result itself.

By emphasizing the dissimilarities, benefits, and barriers of both methods, this comparison also underlines a potential complementarity of HF 2.0 and 3GDP when applied to dialogues or debates over emerging technologies such as nanotechnologies.

Hybrid forums are better equipped to 'act in an uncertain world', to paraphrase the title of Callon, Lascoumes, & Barthe. In other words, and in terms of nanotechnologies, the HF 2.0 would be better adapted to a first approach when the new technology is very not well defined, still uncertain, and when ethical considerations may have to be of general order. HF 2.0 addresses to a larger public and offers a dialogue to an open environment. The 3GDP would be better fitting a deeper perspective aiming at common concertation with concrete hypothesis and results, while HF 2.0 is opened to the unexpected.

A combination of the two methods could be useful when problem-solving and unexpected solutions are expected.

References

Amilien V, Moity-Maïzi P (2019) Controversy and sustainability for geographical indications and localized agro-food systems: thinking about a dynamic link, *British Food Journal*, 121(12), 2981–2994.

Amilien V, Tocco B, Strandbakken P (2019) At the heart of controversies: hybrid forums as an experimental multi-actor tool to enhance sustainable practices in localized agro-food systems, *British Food Journal*, 121(12), 3151–3167.

Callon M, Lascoumes P, Barthe Y (2001) *Agir dans un monde incertain. Essai sur la démocratie technique*, Paris: Seuil.

Callon M, Lascoumes P, Barthe Y (2009). *Acting in an Uncertain World: An Essay on Technical Democracy*, Cambridge, MA: MIT Press.

Farías I (2016) Devising hybrid forums: technical democracy in a dangerous world, *City*, 20(4), 549–562.

Fumel M (2016) *Structure des controverses dans les procédures dialogiques: cas de la mise en place d'un forum hybride en Norvège* (Master in Sciences Sociales Appliquées à l'Alimentation), University of Toulouse, University of Toulouse.

Strandbakken P, Borch A (2015) Third-generation deliberative processes: from consumer research to participatory democracy, in *The Consumer in Society* (Strandbakken, Gronow, eds), Oslo: Abstrakt.

Chapter 16

Conclusion 2020: A More Democratic Science Through Public Engagement?

Pål Strandbakken

SIFO Consumption Research Norway, Oslo Metropolitan University,
Postboks 4, St. Olavs plass 0139 Oslo, Norway
pals@oslomet.no

16.1 Deliberative Processes in Responsible Research and Innovation

As mentioned in Chapter 1, since about 2010, the European buzzword, when it comes to science and technology in society, has been RRI (responsible research and innovation). The term first appeared in 2008, and it has been commonly in use from 2011 (Tancoigne, Randles, & Joly, 2016). It is not altogether clear what RRI 'is' (Rip, 2016), but it has nevertheless become a common point of reference, not the least in the EU program Horizon 2020. There is already a huge literature on RRI (von Schomberg, 2011; Rip, 2014, 2016; Owen, Bessant, & Heintz, 2013; Stilgoe, Owen, & Macnaghten, 2013; Macnaghten, 2016; Timmermans, 2017; European Commission, 2016, to name a

Consumers and Nanotechnology: Deliberative Processes and Methodologies
Edited by Pål Strandbakken, Gerd Scholl, and Eivind Stø
Copyright © 2021 Jenny Stanford Publishing Pte. Ltd.
ISBN 978-981-4877-61-9 (Hardcover), 978-1-003-15985-8 (eBook)
www.jennystanford.com

few), which we will not go into here. Compared to the main theme of this book, the 'traditional' participatory democratic deliberations, RRI is generally directed more at the research agendas, innovation processes, and the laboratories, plus the cutting-edge applications. In addition, however, RRI probably will have to cover the broader issues, the whole set of science and society themes as well.

Behind these and other concepts, the main target or aim has regularly been the need for taking broader social interests and values into account when science develops and when new technologies are introduced. It is about democracy, health and environmental risks, sustainability, fair and equal distribution of risks and benefits, transparency, and so on. In addition, initiatives might have a 'smoother' introduction of controversial technologies as an agenda; following the anti-GMO reactions in Europe, governments and manufacturers are eager not to repeat old mistakes. Most of these initiatives and approaches are aimed at addressing the double challenge of contributing to a better society and better science simultaneously.

Even when the language of science and technology governance changes, some things remain the same. Today, the four dimensions of responsible innovation — anticipation, inclusion, reflexivity, and responsiveness, the so-called AIRR framework (Macnaghten, 2016) — are often highlighted. The deliberations presented in this book have a direct bearing on two of them — inclusion and responsiveness — as well as an indirect bearing on the two others (anticipation and reflexivity).

For deliberative or participatory democracy in science and technology matters to be meaningful, we need to succeed with the two-way dialogue between society on the one hand and researchers and policymakers on the other hand. 'Society' will have to be defined according to the needs of the different dialogues or deliberations; sometimes they will need specific stakeholders, like technology businesses, while other initiatives will call for more 'general stakeholders' such as environmental NGOs and consumer organisations. Most of the deliberations presented here engage citizens, introducing 'society as stakeholder'. A group of citizens might be organised to give an approximation of nationwide representativity (gender, age, geography, educational level, income etc.), or it might be composed of citizens with certain characteristics

— all female groups, senior citizens, consumers, patients, and so on. It all depends on the theme of the deliberation, what it has set out to achieve.

Macnaghten defines, or describes, inclusion as *'associated with the historical decline in the authority of experts, top-down policy making and the deliberative inclusion of new voices in the governance of science and technology'* (Macnaghten, 2016, p. 6). The deliberative processes we have analysed are methods for extracting these new voices and bringing them into the debate and the decision-making.

Inclusion covers activities that involve the identification of stakeholders who are directly and indirectly affected by the R&I process. This dimension acknowledges the presence of different kinds of knowledge, including that of citizens, and it calls for their participation in the design and goal-setting dimensions of a project (Jirotka et al., 2017). So, a wide range of actors and publics should be involved in the entire R&I process, from the start through to the end. This includes their involvement in deliberation and decision-making as a way to create scientific knowledge that is of a higher quality, thanks to the input of a broader range of expertise, disciplines and perspectives (Fossum, Barkved, & Throne-Holst, 2019).

Responsiveness, on the other hand, is about policy's ability to understand and to react to the fears, hopes, and input from society and to translate it into new practice.

Two of the themes we highlighted in the descriptions of the step from second to third generations of deliberative processes (Chapters 13 and 14) are relevant to the responsiveness dimension. First, the problem of rising expectations of participants has been noticed. Here, we mainly insist on establishing clearer links to political processes. Participants in the deliberations should have a realistic understanding of what real-life effects their efforts and engagement might have. We saw an interesting example of such problems when some of the recommendations from the regional Île-de-France deliberations (Chapter 7) missed their target because they addressed things that would have to be dealt with at a national level (military matters, labelling of nanoproducts, regulation of data protection etc.).

Another important aspect of responsiveness is the one where we deal with the relation to representative democracy. When deliberations with citizens somehow simulate a common debate, and stakeholder deliberations resemble hearings, it should be kept in mind that neither debates nor hearings are substitutes for political decision-making, but hopefully they are producing input to the processes.

At another level than the 'dimensions', RRI is supposed to consist of five 'keys': gender, ethics, open science, education to science, and engagement of citizens and civil society in research and innovation. It highlights ethical *'acceptability, sustainability and societal desirability of the innovation process and its marketable products'* (von Schomberg, 2011). Basically, these are a kind of checkpoints for responsible science, mainly aiming at researchers and (science) policymakers in the process of innovation. Gender, ethics, and commitment to open science point to general claims to modern research in open, democratic societies; education to science is mainly beyond the deliberations as we have described them here, even if there is an educational element in many of them. The point we proceed with here is the last one, that is, 'engagement of citizens and civil society in research and innovation', which brings us back to the four dimensions.

From the perspective of the analysed deliberations, the RRI dimensions of inclusion and responsiveness are well covered, while the dimensions of anticipation and reflexivity are more relevant outside the immediate deliberation setting. The first, anticipation, concerns the ability of researchers and organisations *'to develop capacities to ask "what if?" questions'* (Macnaghten, 2016, p. 6), and the second, reflexivity, is about organisations' self-critical monitoring of their own assumptions, the limits to their knowledge, and their awareness of other actors' assumptions and problem framing.

To the extent that RRI is about specific innovations and product development, different versions of the so-called co-creation might be more spot on than the deliberations described and analysed here. Social inclusion through one of the RRI versions of co-creation could work best in innovation processes when the intervention is performed midstream, while the third generation deliberations, even when they strive for specificity, probably work best when applied upstream (Delgado, Kjølberg, & Wickson, 2011). A superficial, but not altogether wrong, way of explaining the differences between the deliberations analysed here and the co-creation approach would be to claim that deliberations are tools for 'politics', while co-creation events are tools for business and researchers. It is about using society to help research become more responsible.

16.2 Third Generation Deliberations and HF 2.0

The subtitle of the book of Callon *et al. Acting in an Uncertain World* is 'An Essay on Technical Democracy' (Callon *et al.*, 2009). The Strength2Food version of 'hybrid forums', called *HF 2.0*, may have been partly inspired by the third generation deliberations, but it operates in a much more unpredictable way when it aims at making technical and scientific matters more democratic (Chapter 15).

While the deliberations resemble a sort of rather codified hearings, aiming at producing text-based input to political processes, HF 2.0 aims more at acting directly in society. At the same time, to a large degree, it allows participants to define their own problems and their own projects. This is perhaps one step more in the empowering of citizen-consumers, and it will likely contribute to reduce a kind of alienation in the lay populace.

This open and more experimental approach might still produce statements (and material for social analysis), but that is really an add-on. It is the common project and the action that is crucial. HF 2.0 is an approach to mobilising local resources and local engagement around controversial issues, and it allows, even encourages, unexpected redefinitions of the situation, like the change in the Norwegian event, when the group went from discussing the utilization of local seafood resources to planning a maritime museum for teaching the public about life in the fjord (Section 15.4, case study 1). The openness of the process might resemble some of the unplanned development we observed in the French 'Nanomonde' cycle of conferences on nanotechnology (Chapter 6):

Recently Vivagora[1] refuses to regard recommendations as the main objective or outcome from their deliberative processes: now they claim that these open debates with large participation instead should aim at helping citizens to formulate better questions about science issues rather than trying to formulate recommendations that in short time have to synthesise contradictory positions and interests. This probably means that decision-makers behind Nanomonde/Vivagora feel that the original design needed to be developed (Section 6.5, Summarising Appraisal).

---

[1]The French NGO organising the cycle

It seems as if Nanomonde would have profited from employing a method more in the hybrid forum vein; that the way this approach deals with disagreement and controversy would have worked better in the very tense French situation. It calls for dealing with citizens' engagement in a less predefined way — more open to transformations and with an aim to use conflicts of interest in a creative way.

The work towards achieving a 'more democratic science through public engagement' will obviously benefit from a wide range of approaches to citizen and stakeholder involvement and not being confined to just one method.

16.3 Participatory and Representative Democracy

To employ the theories and techniques from participatory and deliberative democracy on the themes of science and technology governance is challenging. Neither the local assemblies for taking decisions at the community level (Rousseau-inspired anarchists' solutions), nor the different workers' councils in industry (Pateman, 1977 and others) are directly applicable. This means that participation and deliberation in matters of science and technology has to be framed differently and organised in other ways. As indicated earlier, a lot of these processes are substitutes for debates that fail to appear. This lack of a common, societal debate makes research processes more predictable, hence less interesting and less responsive to societal demands.

Regular democratic theory builds upon the idea of one man/ one vote and is based on the power of the majority. Stakeholder approaches and deliberative processes should supplement the traditional processes by acknowledging and giving legitimacy to lobbying, negotiations, and consensus-driven cooperation, while still being aware that the decisive decisions are taken in the representative systems at municipal, regional, national, or international levels.

In principle, this would also be the case for the organisations that we insist should be regarded as policymakers/political actors. The policies adopted by environmental NGOs, consumer organisations, local ad hoc organisations, and so on might be inspired by statements coming from deliberations in small (and closed?) groups, but formal policy decisions will have to be the responsibility of an elected leadership that is kept accountable by the organisation's members.

To our knowledge, none of the deliberations described here have been used to 'bypass democracy'.

Do we know anything about real influence? Do we have examples of how suggestions and input coming from deliberations have been adopted by and acted on in the representative political system or have been included into the policies of organisations? What about the broader social debate?

We know that the German *Consumer Conference on the Perception of Nanotechnology in the Areas of Food, Cosmetics and Textiles* (Chapter 5) had its results presented at the consumer committee of the German Parliament (Bundestag), at industrial associations, and at the European Food Safety Authority (EFSA), but *'information as to if or to what extent these promotional activities actually had an impact on decision-making is not available though'* (Section 5.4). Political processes take time and the history of specific decisions is hard to trace, but the exposure and dissemination of the results seem promising. Participants in the German event also reported that they brought the topics back into their daily lives and that they *'contributed to an informal social dissemination'*. Any effects of these forms of social dissemination will be hard to trace, however, even if we believe that the aggregate of all such tiny discussions is a society that will be better equipped to deal with the technological change.

The post-event evaluation of the *French Citizens' Conference* (Chapter 7) was a bit different. The group had the opportunity, or the assignment, to influence the region's science policy directly, but the results — according to the arranger — fell between chairs. The general recommendations about investing in basic research and the importance of developing toxicology were probably what the Council would have gone for anyway. Their more specific recommendations, as mentioned, tended to be suggestions for action at the national, not the regional, level. The person in charge of the arrangement, Marc Lipinski, stated 2 years after the event that *'the potential for following up should have been considered more thoroughly in advance'* (Section 7.5).

In a third of our original deliberations, *The Nanotechnology Citizens' Conference in Madison, USA* (Chapter 8), described as having *'most of the characteristics of an ideal deliberative process'*

(Section 8.5), it is stated plainly that the conference *'had no direct impact on policy'*. A similar conclusion was reached at the other US event, that is, the *US National Citizens' Technology Forum on Human Enhancement* (Chapter 9). Here, they conclude that *'it lacked a clear avenue for the outcomes of the process to affect policy and/or research and development trajectories'*. With reference to the responsiveness theme mentioned earlier, here the participants reported a slight *'decrease in their belief that their opinions or actions can actually affect political outcomes'* (Section 9.5). For the American events, we do not know anything about potential long-term effects of raising the themes among persons who might initiate debates in their local communities.

In Chapter 10, under the heading of 'Stakeholder-Oriented Deliberative Processes', three experiments with cross-national deliberative processes were analysed, that is, *The Convergence Seminars*, the *DEMOCS Card Games*, and the *Nanologue Project*. The main conclusion about impact here parallels the citizens' deliberations, but they are worded a bit differently. The processes are understood as *'experimentation with deliberative processes'*, and so *'their primary value is not on their impact on the field of NT policy...'* (Section 10.5). It is possible, however, that the way the impact dimension was defined in the Nanoplat project, focusing on the impact of the initiatives on *'member states NT policy'* is a bit too narrow. At a more diffuse level, the authors mention that an effect might be that they contribute to create a deliberative background for EU's nanotechnology policy: *'Considering the fact that in some areas where the EU level policy will precede those of member states (e.g., in the area of novel food), this impact is not a negligible one'*.

The Nanoplat reporting did not really include the indirect impacts, where stakeholders are regarded as policymakers in the modern governance system; hence, it was not focusing on the importance of influencing such actors' perspectives on nano matters.

When we come to the NanoDiode deliberations, however (Chapter 14), the policy relevance of stakeholders' views was acknowledged, and so the impact aspect was addressed differently. At least two, possibly three, of the deliberations reported some politically relevant influence and effect from the events; others indicated potential indirect impact.

The Norwegian third generation deliberation on Human Enhancement did influence the User Committee on personalized medicine. The User committee later published its position on the subject, partly based on the presentations given. Something similar was reported from the User Committee for Health South East (a patient-based organization) and probably the Older Doctors' Association. The other stakeholders regarded this as an important step in clarifying their own perspectives on the role of social debates and on their respective organizations' role in these processes.

For Germany, the event might have had some indirect political impact. The representatives of the Baden-Wuerttemberg State Ministry for Rural Areas and Consumer Protection, with its own NanoDialogue Baden-Wuerttemberg, indicated that they would take up the results on their Nano-Portal, which might also influence the Ministry's work at the state level and in federal bodies. Further, the RKW Kompetenzzentrum and BG BAU should bring the deliberation experiences forward as stakeholders in different networks and bodies. The political leverage of the professional end user or student participants might be limited, but the awareness that was created would probably lead to discussions on the relevant questions in the different guilds, chambers, and other professional associations.

Politically, the Dutch deliberation was strategically well situated and well attended. Even if the stakeholders felt that the deliberation did not produce any novel insights, it appeared as if they did see it as part of a *continuous societal debate* over questions of nano in food and health issues.

In Austria, the Federal Ministry of Health submitted a statement on nanotechnology initiatives and the importance of the key enabling technology. It brought the outcomes of the discussions into further dialogues within the Austrian Nano Information Commission (NIK), where all nano-relevant ministries were represented. A report was submitted to the ministries, and beyond that, again, political impact might result from increased reflexivity among scientific and business representatives.

In Italy, the 3GDP was considered useful to address the information gap between experts and the general public. Further, it was considered important to increase the nano attention of the players in the food value chain. The wide range of stakeholders involved ensured that different visions and opinions were represented in the debate and stimulated the discussion. Participants were very interested in mapping the views of the different stakeholders participating in the workshop. The event probably was successful in stimulating some reflection on the responsible development of nanotech and fostering a dialogue amongst the various stakeholders. Potentially important for the follow-up phase was the presence at the workshop of journalists. At least a couple of articles on food in the sectorial press were expected, based on the talks and documents presented.

This does not mean that the deliberations under the EU framework programmes have solved the problems of using input from participatory events in regular politics. The deliberations reviewed here are generally rather successful as methodological exercises and as learning platforms for participants, and as such they are important dress rehearsals for more democratic political processes. They are also a kind of inventory of social concerns, hopes, and fears for the emerging nanotechnologies.

They are, however, not very successful as demonstration pieces of the relation between deliberative or participatory democracy, on the one hand, and the regular, representativity based political system, on the other hand. They have been more experiments in participation than they have been real political workshops. All of them, whatever generation, suffer from a lack of clear political commitment. They were, with the possible exception of the Île-de-France events, not given a clear role in decision-making and policy formulation. There was no clear obligation to bring results and statements into the legitimate political processes.

We should perhaps see this 'first wave' of deliberations as a series of experiments; experiments that successfully demonstrated that both citizens and stakeholders are able to provide relevant input to discussions about emerging technologies. When given an introduction to visions, potentials, and risks, they are able to represent the societal perspective(s) and consider ethical, environmental, and even legal aspects of the new challenges. This is without necessarily understanding the physics of it. One metaphor might be societal engagement in traffic solutions: Ordinary people can be given a voice here, even without being able to explain the workings of the combustion engine.

But after this successful wave of deliberations, the challenge for a more democratic science in the future is to 'insert' the results and statements from the deliberative events into political decision-making and regulation. At the moment, the threat to democratic science is more that the deliberations have too little impact on politics than that they are given too much weight compared to the regular system.

It is tempting to reuse the conclusion from the Norwegian-German *Deliberative Processes* project (Stø, Scholl, Strandbakken, & Throne-Holst, 2010):

Deliberative processes represent a democratization of science, and, as long as we distinguish between public discourse and formal decision-making processes, deliberation represent no threat to numerical democracy (p. 3, Executive Summary).

This still holds, and in a way the problem is the opposite: How do we ensure that the results from these events are introduced into the political processes? They contain insights, opinions, and approaches that should not be lost.

References

Callon M, Lascoumes P, Barthe Y (2009). *Acting in an Uncertain World: An Essay on Technical Democracy*, Cambridge, MA: MIT Press.

Delgado A, Kjølberg KL, Wickson F (2011) Public engagement coming of age: from theory to practice in STS encounters with nanotechnology, *Public Understanding of Science*, 20(6), 826–845.

European Commission (2016) http://ec.europa.eu/programmes/horizon2020

Fossum S, Barkved L, Throne-Holst H (2019) Practicing responsible research and innovation in a crowdsourcing project in Norway, *ORBIT Journal*, 2(1), 1–28.

Macnaghten P (2016) *The metis of responsible innovation. Helping society to get better at the conversation between today and tomorrow*, Inaugural lecture, Wageningen University.

Owen R, Bessant J, Heintz M (eds) (2013) *Responsible Innovation: Managing the Responsible Emergence of Science and Innovation in Society*, Chichester: John Wiley & Sons.

Pateman, C (1977) *Participation and Democratic Theory*, Cambridge: Cambridge University Press.

Rip A (2014) The past and future of RRI, *Life Sciences, Society and Policy*, 10(1).

Rip A (2016) The clothes of the emperor: an essay on RRI in and around Brussels, *Journal of Responsible Innovation*, 3(3), 290–304.

von Schomberg R (2011), Prospects for technology assessment in a framework of responsible research and innovation, in *Technikfolgen Abschätzen Lehren: Bildungspotensiale* (Dusseldorf, Beecroft, eds), Wiesbaden: Transdisiplinärer Verlag Methoden.

Stilgoe J, Owen R, Macnaghten P (2013) Developing a framework for responsible innovation, *Research Policy*, 42, 1568–1580.

Stø E, Scholl G, Strandbakken P, Throne-Holst H (2010), *Deliberative processes – increased citizenship or a threat to democracy? A synthesis of recent research within emerging technologies*, SIFO Report 2010, Oslo.

Timmermans J (2017) Mapping the RRI landscape: an overview of organisations, projects, persons, areas and topics, *Responsible Innovation 3: A European Agenda?* (Asveld *et al.*, eds), Springer.

Tancoigne E, Randles S, Joly P-B (2016) Small and divided worlds: a systematic review and scientometric analysis of RRI literature, in *Navigating Towards Shared Responsibility in Research and Innovation. Approach, Process and Results of the Res-AGorA Project* (Lindner, Kuhlmann, Randles, Bested, Gorgoni, Griessler, Loconto, Mejlgaard, eds), Karlsruhe: Fraunhofer ISI.

Authors' Biographies

Michael D. Cobb (born in 1968, USA) is a political scientist currently working as an associate professor at the North Carolina State University. He completed his PhD at the University of Illinois in 2001 and specialises in political psychology and public-opinion research. His main areas of interest include public opinion, political behaviour, and citizen deliberation. E-mail: mdcobb@gw.ncsu.edu

Enikő Demény (born in 1969, Hungary) is a civil engineer currently working as a junior researcher at the Center for Ethics and Law in Biomedicine at the Central European University. She is also a visiting lecturer at Babes Bolayi University in Romania. She holds a PhD in philosophy from 2006. Her main areas of interest include social, legal, ethical, and policy aspects of new converging technologies, and bioethics and social sciences. E-mail: demenye@ceu.hu

Patrick W. Hamlett (born 1946, USA) is an associate professor of political science and of science, technology, and society at North Carolina State University. He earned his PhD at the University of California, Santa Barbara, 1981. His main areas of interest include deliberative democratic theory and science, technology, and public policy. E-mail: patrick.hamlett@gmail.com

François Jégou (born in 1965, France) is a designer currently working as director of the Brussels-based sustainability innovation lab Strategic Design Scenarios, which he founded in 1990. He is also the scientific director of the public innovation lab the 27e Region in France, and teaches strategic design. He graduated from the Ecole Nationale Superieure de Creation Industrielle in Paris, writing about cross-fertilisation between future research and design. His main areas of interest include strategic design, participative scenario building, and new product-services system definition. E-mail: f.jegou@gmail.com

Peter Kakuk (born in 1973, Hungary) is currently working as an assistant professor at the Department of Behavioural Sciences, School of Public Health at the University of Debrecen, Hungary. He holds a PhD in health sciences from the University of Debrecen. His main areas of interest include bioethics and the ethical aspects of health care, normative aspects of the science-society relationship, research ethics, and contemporary challenges on scientific integrity. E-mail: kakukp@dote.hu

Atilla Öner (born in 1955, Turkey) is currently working as an associate professor at the Department of Business Administration at Yeditepe University in Istanbul. He is also the managing director of Management Application and Research Center at Yeditepe University. He holds a PhD in chemical engineering from Yale University. His main areas of interest include manufacturing strategies, technology management, R&D management, foresight, system dynamics modelling, and national innovation systems. E-mail: maoner@yeditepe.edu.tr

Ulrich Petschow (born in 1952, Germany) is an economist currently working as researcher at the Institute for Ecological Economy Research (IÖW) in Berlin. His main areas of interest include environmental policy and governance, innovation and technology, and water and land management. E-mail: ulrich.Petschow@ioew.de

Giampiero Pitisci is an Italian designer who worked with Strategic Design Scenarios (SDS) in Brussels during the Nanoplat project. He is currently working as an independent designer and educator in the design field.

Judit Sandor (born in 1962, Hungary) is a lawyer working currently as a professor of law and political science and since 2005 as the director of the Center for Ethics and Law in Biomedicine (CELAB) at the Central European University in Budapest. She has also been teaching postgraduate level courses on human rights, health care law, bioethics, and bio politics and gender. Her main areas of interest include the ethical, social, and legal implications of new technologies, especially genetic research and assisted reproduction. E-mail: sandorj@ceu.hu

Özcan Saritas (born in 1976, Turkey) is currently working as a research fellow at the Manchester Institute of Innovation Research,

Manchester Business School, United Kingdom, and is the editor of *Foresight: The journal of future studies, strategic thinking, and policy*. He holds a PhD entitled *System Thinking for Foresight* from the Manchester Business School, PREST Research Centre. His main areas of interest include long-term policy and strategy-making with particular emphasis upon Foresight methodologies and their implementation in socio-economic and technological fields at the supra-national, national, regional and sectorial levels. E-mail: ozcan.saritas@mbs.ac.uk

Gerd Scholl (born in 1966, Germany) is an economist and an independent consultant. Formerly, he worked as a senior researcher at the Institute for Ecological Economy Research (IÖW) in Berlin. He has a degree in economics from Bonn University, Germany, and completed his PhD in marketing from University of Oldenberg, Germany, in 2008. His main areas of interest include sustainable consumption and production, sustainable marketing, and consumers and new technologies.

Eivind Stø (born in 1945, Norway) is a political scientist, and he retired as a research director from the National Institute for Consumer Research in Norway. His main areas of interest include the sociology of consumption, sustainable production and consumption, and consumer complaints.

Pål Strandbakken (born in 1957, Norway) is a sociologist currently working as a researcher at SIFO Consumption Research Norway, Oslo Metropolitan University, Norway. He has a degree in sociology from the University of Oslo, Norway, in 1987 on Weber's Protestant Ethic thesis and holds a PhD in sociology from the University of Tromsø, Norway, on product durability and the environment. His main areas of interest include environmental sociology, consumer studies, and science and technology studies. E-mail: pals@oslomet.no

Harald Throne-Holst (born in 1970, Norway) has an MSc in inorganic chemistry and is currently working as a researcher at SIFO Consumption Research Norway, Oslo Metropolitan University, Norway. His main areas of interest include ethical and societal aspects of nanotechnology, energy use in households, and consumption and the environment. E-mail: harth@oslomet.no

Fern Wickson (born in 1976, Australia) is a cross-disciplinary researcher currently working at GenØk Centre for Biosafety in Tromsø. She holds a PhD on Australia's environmental regulation of genetically modified crops from the University of Wollongong. She has also worked as a postdoctoral researcher at the Centre for the Study of the Sciences and the Humanities at the University of Bergen. Her main areas of interest include environmental philosophy, ecotoxicology, the politics of risk and uncertainty, and the governance of emerging technologies.
E-mail: fern.wickson@uit.no

Virginie Amilien (born in 1964, France) is a Research Professor in food culture at SIFO Consumption Research Norway at the Oslo Metropolitan University, Norway and she holds a PhD in Scandinavian literature and languages. Her main fields of research are food culture, culture of consumption, food practices, local food, national identity and tourism.
E-mail: viram@oslomet.no

Barbara Tocco (born in 1987, Italy), holds a PhD in economics and currently holds a position as Research Associate, Newcastle University Business School, Newcastle upon Tyne, UK. Her main fields of research are agri-food supply chains, food sustainability, rural development and policy analysis: sustainable food production-consumption, food consumer behaviour, consumer perceptions, farmers'/managers' perceptions and decision-making, 'alternative' food systems, short food supply chains, local and community food initiatives, food cultures, participatory research methods and mixed-methods research.
E-mail: barbara.tocco@newcastle.ac.uk

Index

Printed in the United States
by Baker & Taylor Publisher Services

Printed in the United States
by Baker & Taylor Publisher Services